IUFRO Research Series

The International Union of Forest Research Organizations (IUFRO), with its 14,000 scientists from 700 member institutions in over 100 countries, is organized into nearly 300 research units that annually hold approximately 80 conferences, workshops and other meetings. The individual papers, proceedings and other material arising from these units and meetings are often published but in a wide array of different journals and other publications. The object of the IUFRO Research Series is to offer a single, uniform outlet for high-quality publications arising from major IUFRO meetings and other products of IUFRO's research units.

The editing, publishing and dissemination experience of CABI Publishing and the huge spread of scientific endeavours of IUFRO combine here to make information widely available that is of value to policy makers, resource managers, peer scientists and educators. The Board of IUFRO forms the Editorial Advisory Board for the series and provides the monitoring and uniformity that such a high-quality series requires in addition to the editorial work of the conference organizers.

While adding a new body of information to the plethora currently dealing with forestry and related resources, this series seeks to provide a single, uniform forum and style that all forest scientists will turn to first as an outlet for their conference material and other products, and that the users of information will also see as a reliable and reputable source.

Although the official languages of IUFRO include English, French, German and Spanish, the majority of modern scientific papers are published in English. In this series, all books will be published in English as the main language, allowing papers occasionally to be in other languages. Guidelines for submitting and publishing material in this series are available from the Publisher, Books and Reference Works, CABI Publishing, CAB International, Wallingford, Oxfordshire OX10 8DE, UK, and the IUFRO Secretariat, Hauptstrasse 7, A-1140 Vienna – Hadersorf, Austria.

IUFRO Research Series

Titles available:

Forest Biodiversity: Lessons from History for Conservation

Edited by

O. Honnay,
K. Verheyen,
B. Bossuyt

and

M. Hermy
Katholieke Universiteit Leuven, Faculty of Applied Biological Sciences, Laboratory for Forest, Nature and Landscape Research, Leuven, Belgium

CABI Publishing
in association with
The International Union of Forestry Research Organizations
(IUFRO)

CABI Publishing is a division of CAB International

CABI Publishing
CAB International
Wallingford
Oxfordshire OX10 8DE
UK

CABI Publishing
875 Massachusetts Avenue
7th Floor
Cambridge, MA 02139
USA

Tel: +44 (0)1491 832111
Fax: +44 (0)1491 833508
Email: cabi@cabi.org
Website: www.cabi-publishing.org

Tel: +1 617 395 4056
Fax: +1 617 354 6875
Email: cabi-nao@cabi.org

A catalogue record for this book is available from the British Library, London, UK.

Library of Congress Cataloging-in-Publication Data
Forest Biodiversity : lessons from history for conservation / edited by
O. Honnay . . . [et al.].
    p.   cm.
Includes bibliographical references (p.).
  ISBN 0-85199-802-X (alk. paper)
  1. Forest conservation--Congresses.   2. Forest ecology--Congresses.
3. Plant diversity conservation--Congresses.   I. Honnay, O. (Olivier)
II. Title.
  SD411.F57 2004
  333.75'16--dc22

2003016396

ISBN 0 85199 802 X

Typeset by AMA DataSet Ltd, UK.
Printed and bound in the UK by Cromwell Press, Trowbridge.

Contents

Contributors

K.N.A. Alexander, *Ancient Tree Forum, c/o Woodland Trust, Autumn Park, Grantham, Lincolnshire NG31 6LL, UK*

G.B. Blank, *Department of Forestry, North Carolina State University, Box 8002, Raleigh, NC 27695-8002, USA*

B. Bossuyt, *Laboratory for Forest, Nature and Landscape Research, University of Leuven, Vital Decosterstraat 102, B-3000 Leuven, Belgium*

R.H.W. Bradshaw, *Environmental History Research Group, Geological Survey of Denmark and Greenland, Øster Voldgade 10, DK-1350, Copenhagen, Denmark*

J. Brunet, *Southern Swedish Forest Research Centre, Swedish University of Agricultural Sciences, Box 49, SE-230 53 Alnarp, Sweden*

J.E. Butler, *Ancient Tree Forum, c/o Woodland Trust, Autumn Park, Grantham, Lincolnshire NG31 6LL, UK*

P. Coquillard, *Equipe Gestion de la Biodiversité EA3156, Université de Nice, Faculté des Sciences, Parc Valrose, 06108, Nice Cedex 2, France*

T. Curt, *Cemagref, U.R., Dynamiques et Fonctions des Espaces Ruraux, Campus des Cézeaux, B. P. F-50085, 63172 Aubière Cedex, France*

E. Dambrine, *INRA, Equipe Cycles Biogéochimiques, 54280 Seichamps, France*

K. Desender, *Department Entomology, RBINSc, Vautierstraat 29, B-1000, Brussels, Belgium*

B. De Vos, *Institute for Forestry and Game Management, AMINAL, Gaverstraat 4, B-9500 Geraardsbergen, Belgium*

H. Dhuyvetter, *Department Entomology, RBINSc, Vautierstraat 29, B-1000, Brussels, Belgium*

K.M. Flinn, *Department of Ecology and Evolutionary Biology, Corson Hall, Cornell University, Ithaca, NY 14853-2701, USA*

E. Gaublomme, *Department Entomology, RBINSc, Vautierstraat 29, B-1000, Brussels, Belgium*

F. Gosselin, *Cemagref – Agricultural and Environmental Engineering Research, Domaine des Barres, F-45290 Nogent sur Vernisson, France*

G.E. Hannon, *Southern Swedish Forest Research Centre, Box 49, S-230 53, Alnarp, Sweden*

W. Härdtle, *Institute of Ecology, University of Lüneburg, 21332 Lüneburg, Germany*

M. Hermy, *Laboratory for Forest, Nature and Landscape Research, University of Leuven, Vital Decosterstraat 102, B-3000 Leuven, Belgium*

O. Honnay, *Laboratory for Forest, Nature and Landscape Research, University of Leuven, Vital Decosterstraat 102, B-3000 Leuven, Belgium*

R. Howard, *University Tree-ring Dating Laboratory, Department of Archaeology, University of Nottingham NG7 2RD, UK*

J.W. Humphrey, *Forest Research, Northern Research Station, Roslin, Midlothian, EH25 9SY, UK*

E. Johann, *University of Agricultural Sciences Vienna, Wlassakstrasse 56, A-1130 Vienna, Austria*

M.R. Jukes, *Forest Research, Alice Holt Research Station, Wrecclesham, Farnham, Surrey, GU10 4LH, UK*

K.J. Kirby, *English Nature, Northminster House, Peterborough, PE1 1UA, UK*

C. Lavers, *School of Geography, University of Nottingham, Nottingham NG7 2RD, UK*

J.E. Lawesson, *Institute of Biological Sciences, Aarhus University, Nordlandsvej 68, DK-8240 Risskov, Denmark*

J. Lhonoré, *Cemagref – Agricultural and Environmental Engineering Research, Domaine des Barres, F-45290 Nogent sur Vernisson, France*

P.L. Marks, *Department of Ecology and Evolutionary Biology, Corson Hall, Cornell University, Ithaca, NY 14853, USA*

M.J. Mortimer, *Virginia Polytechnic Institute, State University, Department of Forestry, Blackburg, VA 24061, USA*

L. Östlund, *Department of Forest Vegetation Ecology, Swedish University of Agricultural Sciences, 901 83 Umeå, Sweden*

A.J. Peace, *Forest Research, Alice Holt Research Station, Wrecclesham, Farnham, Surrey GU10 4LH, UK*

E.L. Poulsom, *Forest Research, Northern Research Station, Roslin, Midlothian, EH25 9SY, UK*

B. Prévosto, *Cemagref, U.R., Dynamiques et Fonctions des Espaces Ruraux, Campus des Cézeaux, B.P. F-50085, 63 172 Aubière Cedex, France*

E. Richard, *Cemagref – Agricultural and Environmental Engineering Research, Domaine des Barres, F-45290 Nogent sur Vernisson, France*

M. Vellend, *Department of Ecology and Evolutionary Biology, Cornell University, Ithaca, NY 14853, USA*

P. Verdyck, *Department Entomology, RBINSc, Vautierstraat 29, B-1000, Brussels, Belgium and University Antwerp, Groenenborgerlaan, 171, B-2020, Antwerp, Belgium*

K. Verheyen, *Laboratory for Forest, Nature and Landscape Research, University of Leuven, Vital Decosterstraat 102, B-3000 Leuven, Belgium*

G. von Oheimb, *Institute of Ecology, University of Lüneburg, 21332 Lüneburg, Germany*

C. Watkins, *School of Geography, University of Nottingham, Nottingham NG7 2RD, UK*

C. Westphal, *Institute of Ecology, University of Lüneburg, 21332 Lüneburg, Germany*

M. Wulf, *ZALF e. V. Müncheberg, Eberswalder Strasse 84, 15374 Müncheberg, Germany*

Preface

All the chapters in this book are based on presentations given at the symposium 'History and Forest Biodiversity: Challenges for Conservation' which was held in Leuven, Belgium in January 2003. The symposium was a joint initiative of the Laboratory of Forest, Nature and Landscape Research of the University of Leuven and the IUFRO (International Union of Forest Research Organizations) units on 'Ecological History' (6.07.04) and 'Biodiversity' (8.07.00). The Forest and Green Spaces Division of the Ministry of Flemish Community kindly sponsored this event. Over 130 people from 20 different countries attended the symposium, in which 51 lectures and 45 posters were presented. Most participants came from north-western Europe, but significant delegations from the USA and Eastern Europe were also present.

We would like to thank our colleagues at the Laboratory of Forest, Nature and Landscape Research for providing secretarial and organizational assistance, especially Sofie Bruneel, Pieter Roovers and Patrick Endels, as well as Tim Hardwick of CAB International for his assistance with the publication of this book.

Olivier Honnay, Kris Verheyen, Beatrijs Bossuyt and Martin Hermy
Leuven, July 2003

What History Can Teach us About Present and Future Forest Biodiversity   1

K. Verheyen, O. Honnay, B. Bossuyt and M. Hermy

Laboratory of Forest, Nature and Landscape Research, University of Leuven, Leuven, Belgium

1.1 Introduction

Reconstructing the past to evaluate its effects on present ecosystems is like a puzzle, where the researcher knows the outcome but not the rules. It requires the integration of a wealth of diverse information. Each kind of record provides a unique type of information, each with its own spatial and temporal scale. By integrating these sources, a historical ecologist can piece together a picture of past activities and communities in order to formulate and test hypotheses about causes of past changes and the contributions of past processes to present ecosystems and landscapes.

This phrase from Emily Russell's book (1997) *People and Land Through Time. Linking Ecology and History* elegantly introduces the purpose and theme of this book, in which an attempt is made to give an overview of current knowledge about the legacies of the past and implications for present-day biodiversity in forest ecosystems.

Interest in the ecological history of forests is not new. Several decades ago, pioneers such as Oliver Rackham (1976, 1980) and George Peterken (1974) linked a thorough knowledge of forest history to the present-day structure and composition of forests in general, and their constituent trees in particular. Their work has triggered a new era of interdisciplinary research on forest history, utilizing both cultural evidence, such as written records and maps, and biological evidence, such as contemporary vegetation surveys and data from the sedimentary record (pollen, charcoal, etc.). The proceedings of two previous IUFRO symposia dedicated to this topic provided a good overview of the current status of this field of research (see Kirby and Watkins, 1998; Watkins, 1998; Agnoletti and Anderson, 2000a,b).

Following this growing awareness of the long-term and often pervasive human impact on forest ecosystems, questions have been raised about the consequences of past human impact for forest biodiversity (e.g. Peterken, 1981). Increasingly, it is being realized that the recovery from past (human) disturbances is a very slow process which may take up to hundreds of years, and sometimes systems may never return to their pre-disturbance state. The recent synthesis for the New England forests, published as a special issue of *Journal of Biogeography* (Foster, 2002), can serve as an excellent example in this respect.

Given these insights, the main body of research presented in this book focuses on the diverse impact of forest history in general, and of forest continuity, fragmentation and past management in particular, on the diversity and distribution of species. In addition, some exemplary studies of Holocene forest history are given in the first chapters, to introduce the reader to the diversity of historical events that should be considered. In turn, the final chapters in this book deal with the implications for conservation of a historical component underlying patterns of biodiversity, and the way sound management policies should be implemented. Finally, it should be noted that the vast majority of the research presented concerns temperate forests in Europe and North America. This does not mean that the topic is not relevant in other regions or biomes, it merely indicates that more research is needed in those areas.

1.2 Some Patterns in the Holocene History of Temperate Forests

Nearly all species present in the temperate forests have (re-)colonized from their refugia during the post-glacial Holocene period. Although scientific consensus exists on the location of the main refugia and on the species-specific nature of the subsequent colonization patterns (Hewitt, 2000), other topics remain controversial. For example, Frans Vera (2000) recently triggered a vigorous debate on the nature of the original 'wildwoods', by arguing that they were not closed-canopy forests (e.g. Peterken, 1996), but park-like, open landscapes maintained by large herbivores such as horses, auroch and bison. It seems that the scientific community is currently split between 'believers' and 'non-believers' in this topic. Probably a more nuanced view, in which openness is not only the result of the action of large herbivores, but of an interaction with several other disturbance regimes, such as wind and fire, and with natural site conditions, seems more realistic (Svenning, 2002; Lawesson, 2003; Bradshaw and Hannon, Chapter 2, this volume). The use of this theory as a justification for the re-introduction of large herbivores in forests of all kinds seems therefore somewhat premature. Emulation of natural dynamics in modified ecosystems will continue to be a key research topic (Angermeier, 2000).

Since human settlement and the development of agriculture, several periods of forest regression and expansion have occurred, but their timing and the extent of land-use change vary from one region to another. For instance, in most of north-western Europe, the Roman period was probably the first period of large-scale forest destruction. However, forests recovered to a large extent during the 'Dark Ages' that followed the collapse of the Roman Empire. Subsequent settlement and forest reclamation patterns during the Middle Ages and later periods occurred, at least partly, independently of those during Roman periods (Dupouey *et al.*, 2002; Van Walleghem *et al.*, 2003). In contrast, large-scale reclamation in north-eastern America only started with the arrival of European colonists, in the 17th and 18th centuries, and peaked in the middle of the 19th century. At present, forests have recovered spontaneously to a large extent, due to the shift of agriculture to more suitable agricultural lands in the Great Plains (Whitney, 1994). A similar process is currently occurring in parts of Europe that are also less suitable for agriculture, as exemplified by Johann (Chapter 3, this volume) and Prévosto *et al.* (Chapter 4, this volume) for mountainous parts of Austria and France, respectively.

However, especially in densely populated areas such as north-western Europe, forests that were not cleared for agriculture are, of course, not at all undisturbed remnants of primary forest. The majority of forests have been managed over a long time period for the resources they provide (wood, litter, fruits, fodder, game, etc.), which has often completely changed their structure and composition (e.g. Rackham, 1980; Tack *et al.*, 1993). One need only think of the widespread coppicing practices. Since the industrial and agricultural revolutions, forests served almost entirely as a source for large timber, while other uses, such as coppicing and grazing, were abandoned. The impact of the latter evolution on the structure and dynamics of Swedish forests is illustrated by Östlund (Chapter 5, this volume).

To conclude, it appears that the landscape we have inherited from our ancestors consists of forest patches of different age and origin that have been subjected to different intensities of management. It is evident that this must have an impact on their biodiversity.

1.3 Impact of Forest History on Biodiversity

The different impacts on forest ecosystems can be ranked according to their intensity. Management practices such as single-tree extraction can be considered to be low-intensity impacts, while the complete transformation of forest into, for instance, agricultural land is situated at the other end of the scale. Most of the work included in this book deals with the consequences of the latter impact (Chapters 6–12), although two chapters are devoted to the impacts of management practices (Chapters 13 and 14).

In general, forest species do not survive periods of agricultural use and, therefore, have to recolonize the site when it is reforested. At least for vascular plant species, it has been widely observed that such a recovery may take tens to hundreds of years (e.g. Peterken and Game, 1984; Whitney and Foster, 1988; Dzwonko and Loster, 1990; Lawesson *et al.*, 1998; Hermy *et al.*, 1999; Bellemare *et al.*, 2002). In general, qualitative (i.e. species identities) but no quantitative (i.e. species number) differences between ancient and recent forests are reported. Wulf (Chapter 6), Flinn and Marks (Chapter 7) and Lawesson (Chapter 8) further document this for plant species, while Desender *et al.* (Chapter 9) find similar patterns for the ground beetle, *Carabus problematicus*.

Recolonization depends on the availability of suitable habitat, on the one hand, and on the mobility of the species in question, on the other. By definition, suitable habitat is taxon specific. For forest plants, habitat is generally suitable as soon as the canopy closes, although past agricultural use can have a severe, and sometimes irreversible, impact on the site conditions (e.g. Koerner *et al.*, 1997; Bossuyt *et al.*, 1999a; Verheyen *et al.*, 1999). One of the most dramatic examples in this respect is given by Dupouey *et al.* (2002), who demonstrated persistent effects of Roman agriculture on present-day soil and vegetation characteristics in what had been considered 'ancient forest', based on historical documents from recent centuries. By contrast, for other taxonomic groups, such as fungi, lichens, saproxylic beetles, etc., that often depend on veteran trees, habitat suitability sometimes takes hundreds of years to develop. The species mobility, and hence its colonization capacity, depends both on characteristics of the landscape and on life-history traits of the species. It has been shown that many typical forest species have a low dispersal capacity, or produce few offspring or diaspores (for plants see, for instance, Verheyen *et al.*, 2003; and for beetles, Ranius, 2000). Likewise, spatial isolation from colonization sources has been identified regularly as an important factor explaining colonization patterns (e.g. Matlack, 1994; Brunet and von Oheimb, 1998; Bossuyt *et al.*, 1999b; Butaye *et al.*, 2001). However, far less attention has been paid to the effects of population size in the colonization sources and to the effects of the so-called 'adjacency arrangement' *sensu* Hersperger and Forman (2003) of the target patch. Brunet (Chapter 10) demonstrate that most of the variance in forest understorey richness in recent forests can be explained by their population size in adjacent colonization sources, while Humphrey *et al.* (Chapter 11) found that the species numbers of bryophytes and ground beetles that had colonized recent forest plantations increased with the fraction of semi-natural habitat in the surrounding landscape. Building upon findings that dispersal limitation is the dominant force regulating the recovery of forest herbs in recent forests, Vellend (Chapter 12) developed a simple metapopulation model that enabled him to estimate the time course needed for the recovery of forest plant diversity in fragmented landscapes. Both model and empirical results indicate that, in particular, the extent of ancient forest loss has a critical influence on the rate of recovery. A steep rise in recovery time (up to several hundreds of years) is observed when the proportion of ancient forest in the landscape decreases below 10–15%.

Recovery following management practices is generally faster (but see Duffy and Meier, 1992), since it is more likely that species survive locally or that nearby populations are still present which can serve as colonization sources. The effects of management therefore more often relate to changes in relative abundances, but less so to species composition. This is demonstrated by Richard *et al.* (Chapter 13), who studied the effects of regeneration felling on ground beetle communities, and by Kirby (Chapter 14), who found changes in the relative cover of the ground flora change due to intense deer grazing.

Now, how should we use this knowledge to build a strategy for conservation?

1.4 Using Forest History to Set Priorities for Conservation

A crucial step in any protection scheme is to identify what needs to be conserved. Peterken (1996) argued that ancient forest or, more generally, habitat continuity should remain the priority, but that we should move from a strict historical basis of forest nature conservation (e.g. the continuation of coppicing) to a strategy in which more natural processes and structures are allowed. Along the same lines, Angermeier (2000) recently argued that naturalness is the most reasonable guide in setting conservation goals and priorities. Reliable assessments of naturalness are therefore needed. Such an attempt is made by Wesphal *et al.* (Chapter 15) who explicitly take the historical development of the site into account to assess its naturalness. An understanding of the naturalness and history of a site are also crucial determining references for ecological restoration (cf. Egan and Howell, 2001). An example dealing with the restoration of now-threatened longleaf pine (*Pinus palustris*) communities is given by Blank in Chapter 16.

Networks of reserves are critical to conservation, but conservation should be more than partitioning ecosystems between reserves and areas of intensive human use. A landscape-scale perspective on forest conservation is therefore needed (cf. Peterken, 1996). Forest-like habitats outside forests, such as hedges, hollow roads, individual trees, etc., also deserve protection, especially when they have a long continuity. The plea of Alexander and Butler (Chapter 17) for the value of ancient trees as habitat for many species lost from true forest habitat, therefore deserves attention. The approach by Watkins *et al.* (Chapter 18) to evaluate dead wood in ancient trees provides us with a basis for their management.

Also, from a social perspective, Angermeier (2000) concluded that reserves alone are not sufficient to conserve biodiversity, since they often aggravate society's ecological disconnectedness and trivialize the role of

natural ecosystems in sustaining society. Conservationists should help to
re-establish the connection of society to ecosystems by promoting recognition
of a broad array of ecological values and their relation to natural biotas. However, by reviewing the US legislation for conserving forest biodiversity over the
past 300 years, Mortimer (Chapter 19) concludes that there is still a long
way to go to convince landowners of the merits of (forest) biodiversity. Future
conservation programmes should therefore be coordinated at regional scales
and should include 'working' landscapes as integral components.

1.5 Conclusions

To conclude, we think that this book is illustrative of the evolution in forest
historical ecological research over the past two or three decades. While
ecologists initially focused on relationships between species patterns and forest
history, and attempted to find biological indicators for forest continuity, more
and more emphasis is being put on the processes behind these patterns. Thus,
unlike the recent criticisms formulated by Norden and Appelqvist (2001) and
Rolstad *et al.* (2002), the field of research has moved on to find causal relationships between forest history and species distribution patterns. The way to proceed is captured nicely in the following phrase by Rolstad *et al.* (2002): 'Instead
of using presumably slow-dispersing species to indicate the likely history of a
forest stand, we should use forest stands of known history to help us understand the dispersal ecology of their threatened inhabitants'. Indeed, we should
try to use forest history to learn more about the ecology of the species and, for
instance, attempt to distinguish between species that are mainly limited by
habitat quality (e.g. certain structural components) and species that have poor
dispersal ability. To achieve these goals, integrated research is needed, consisting of the following three steps (Rolstad *et al.*, 2002). First, the forest history
should be reconstructed in great detail. Contrary to Norden and Appelqvist
(2001), who stated that 'too much effort is being spent on the analysis of archi-
val material and the reconstruction of old landscapes', we think that ecologists
should put more effort into this topic, thereby adopting an interdisciplinary
approach to extend maximally the resolution and the time period studied
(Foster *et al.*, 1996). More cooperation between plant and animal ecologists, on
the one hand, and palaeo- and dendroecologists, historians, soil scientists, etc.,
on the other, is badly needed. Next, the analysis of species distribution patterns
across forests differing in well-defined historical features, but constant in other
environmental factors, will allow the development of hypotheses about the
key factors explaining their distribution. In a last step, controlled experiments
should be used to test these hypotheses. While a lot of work has already been
done on vascular plants and some groups of invertebrates, other groups, such
as fungi and epiphytes, should not be neglected. Ultimately, all this should
lead to a mechanistic understanding of the factors affecting forest biodiversity

which, in turn, should enable us to give adequate guidelines to managers and to persuade policy-makers.

Acknowledgements

The authors wish to thank Etienne Branquart, Jonas Lawesson, Jonathan Humphrey and Jörg Brunet for their comments on earlier drafts of this chapter.

References

Agnoletti, M. and Anderson, S. (eds) (2000a) *Methods and Approaches in Forest History*. IUFRO Research Series. CAB International, Wallingford, UK.

Agnoletti, M. and Anderson, S. (eds) (2000b) *Forest History: International Studies on Socioeconomic and Forest Ecosystem Change*. IUFRO Research Series. CAB International, Wallingford, UK.

Angermeier, P.L. (2000) The natural imperative for biological conservation. *Conservation Biology* 14, 373–381.

Bellemare, J., Motzkin, G. and Foster, D.R. (2002) Legacies of the agricultural past in the forested present: an assessment of historical land-use effects on rich mesic forests. *Journal of Biogeography* 29, 1401–1420.

Bossuyt, B., Deckers, J. and Hermy, M. (1999a) A field methodology for assessing man-made disturbance in forest soils developed in loess. *Soil Use and Management* 15, 14–20.

Bossuyt, B., Hermy, M. and Deckers, J. (1999b) Migration of herbaceous plant species across ancient–recent forest ecotones in central Belgium. *Journal of Ecology* 87, 628–638.

Brunet, J. and von Oheimb, G. (1998) Migration of vascular plants to secondary woodlands in southern Sweden. *Journal of Ecology* 86, 429–438.

Butaye, J., Jacquemyn, H. and Hermy, M. (2001) Differential colonization causing non-random forest plant community structure in a fragmented agricultural landscape. *Ecography* 24, 369–380.

Duffy, D.C. and Meier, A.J. (1992) Do Appalachian herbaceous understoreys ever recover from clearcutting? *Conservation Biology* 6, 196–200.

Dupouey, J.L., Dambrine, E., Laffite, J.D. and Moares, C. (2002) Irreversible impact of past land use on forest soils and biodiversity. *Ecology* 83, 2978–2984.

Dzwonko, Z. and Loster, S. (1990) Vegetation differentation and secondary succession on a limestone hill in southern Poland. *Journal of Vegetation Science* 1, 615–622.

Egan, D. and Howell, E. (2001) *The Historical Ecology Handbook: a Restorationist's Guide to Reference Ecosystems*. Island Press, Washington, DC.

Foster, D.R. (2002) Insights from historical geography to ecology and conservation: lessons from the New England landscape. *Journal of Biogeography* 29, 1269–1589.

Foster, D.R., Orwig, D.A. and McLachlan, J.S. (1996) Ecological and conservation insights from reconstructive studies of temperate old-growth forests. *Trends in Ecology and Evolution* 11, 419–424.

Hermy, M., Honnay, O., Firbank, L., Grashof-Bokdam, C.J. and Lawesson, J.E. (1999) An ecological comparison between ancient and other forest plant species of Europe, and the implications for forest conservation. *Biological Conservation* 91, 9–22.

Hersperger, A.M. and Forman, R.T.T. (2003) Adjacency arrangement effects on plant diversity and composition in woodland patches. *Oikos* 101, 279–290.

Hewitt, G. (2000) The genetic legacy of the Quaternary ice ages. *Nature* 405, 907–913.

Kirby, K.J. and Watkins, C. (eds) (1998) *The Ecological History of European Forests.* CAB International, Wallingford, UK.

Koerner, W., Dupouey, J.L., Dambrine, E. and Benoît, M. (1997) Influence of past land use on the vegetation and soils of present day forest in the Vosges mountains, France. *Journal of Ecology* 85, 351–358.

Lawesson, J.E. (2003) Review of grazing ecology and forest history by F.W.M. Vera. *Journal of Vegetation Science* 14, 137–140.

Lawesson, J.E., de Blust, G., Grashof, C., Firbank, L., Honnay, O., Hermy, M., Hobitz, P. and Jensen, L.M. (1998) Species diversity and area-relationships in Danish beech forests. *Forest Ecology and Management* 106, 235–245.

Matlack, G.R. (1994) Plant species migration in a mixed-history forest landscape in eastern North America. *Ecology* 75, 1491–1502.

Norden, B. and Appelqvist, T. (2001) Conceptual problems of ecological continuity and its bioindicators. *Biodiversity and Conservation* 10, 779–791.

Peterken, G.F. (1974) A method for assessing woodland flora conservation using indicator species. *Biological Conservation* 6, 239–245.

Peterken, G.F. (1981) *Woodland Conservation and Management.* Chapman & Hall, London.

Peterken, G.F. (1996) *Natural Woodland.* Cambridge University Press, Cambridge.

Peterken, G.F. and Game, M. (1984). Historical factors affecting the number and distribution of vascular plant species in the woodlands of central Lincolnshire. *Journal of Ecology* 72, 155–182.

Rackham, O. (1976) *Trees and Woodland in the British Landscape.* Dent, London.

Rackham, O. (1980) *Ancient Woodland: its History, Vegetation and Uses in England.* Edward Arnold, London.

Ranius, T. (2000) Minimum viable metapopulation size of a beetle, *Osmoderma eremita*, living in tree hollows. *Animal Conservation* 3, 37–43.

Rolstad, J., Gjerde, I., Gundersen, V.S. and Saetersdal, M. (2002) Use of indicator species to assess forest continuity: a critique. *Conservation Biology* 16, 253–257.

Russell, E. (1997) *People and Land Through Time. Linking Ecology and History.* Yale University Press, New Haven, Connecticut.

Svenning, J.C. (2002) A review of natural vegetation openness in north-western Europe. *Biological Conservation* 104, 133–148.

Tack, G., Van den Brempt, P. and Hermy, M. (1993) *Bossen van Vlaanderen. Een Historische Ecologie.* Davidsfonds, Leuven, Belgium.

Van Walleghem, T., Van den Eeckhaut, M., Poesen, J., Deckers, J., Nachtergaele, J., Van Oost, K. and Slenters, C. (2003) Characteristics and controlling factors of old gullies under forest in a temperate humid climate: a case study from the Meerdaal Forest (Central Belgium). *Geomorphology* 1333, 1–15.

Vera, F.W.M. (2000) *Grazing Ecology and Forest History*. CAB International, Wallingford, UK.

Verheyen, K., Bossuyt, B., Hermy, M. and Tack, G. (1999) The land use history (1278–1990) of a mixed hardwood forest in central Belgium and its relationship with chemical soil characteristics. *Journal of Biogeography* 26, 1115–1128.

Verheyen, K., Honnay, O., Motzkin, G., Hermy, M. and Foster, D.R. (2003) Response of forest plant species to land-use change: a life-history trait-based approach. *Journal of Ecology* 91, 563–577.

Watkins, C. (ed.) (1998) *European Woods and Forests: Studies in Cultural History*. CAB International, Wallingford, UK.

Whitney, G.G. (1994) *From Coastal Wilderness to Fruited Plain*. Cambridge University Press, New York.

Whitney, G.G. and Foster, D.R. (1988) Overstorey composition and age as determinants of the understorey flora of woods of Central New England. *Journal of Ecology* 76, 867–876.

The Holocene Structure of North-west European Temperate Forest Induced from Palaeoecological Data

2

R.H.W. Bradshaw[1] and G.E. Hannon[2]

[1]Environmental History Research Group, Geological Survey of Denmark and Greenland, Copenhagen, Denmark; [2]Southern Swedish Forest Research Centre, Alnarp, Sweden

Several threatened forest species are currently associated with semi-open, 'wood–pasture' conditions and do not thrive in present-day, non-intervention temperate forests of north-western Europe. We assess the changing importance through time of five disturbance agencies that open forest canopies and induce the past structure of these forests. The influence of browsing and grazing animals has varied through time, but was most intense from domestic animals during recent centuries. The role of large ungulates on forest structure during the early Holocene was negligible. Fires of both natural and anthropogenic origin have been of importance in the past, but have now virtually ceased. Past effects of waterlogging have been severely reduced by drainage schemes. Wind-throw has been a relatively constant factor through time, while anthropogenic influence has dominated forest structure, particularly during recent centuries. 'Natural' forest structure is probably more open and varied than found in present-day, non-intervention, reference forests, due to variable combinations of these disturbance agencies.

2.1 Introduction

Forest structure, particularly its homogeneity and degree of openness, is one key determinant of forest biodiversity, and many red-listed species of insects, bryophytes and lichens are associated with sun-exposed trunks (Rose, 1976; Nilsson, 1997). The degree of openness and the proportion of large gaps and forest glades are also important for the regeneration of light-demanding forest trees, such as *Quercus* and *Corylus*, which have been abundant in many parts of north-western Europe throughout the Holocene (Berglund *et al.*, 1996; Vera, 2000). These observations suggest that forests must have contained a significant proportion of open area in the past, yet unmanaged 'old growth' forest in much of Europe today tends to be rather dense and dark (Peterken, 1996; Svenning, 2002). Populations of *Quercus* and *Corylus* are often reducing in size in unmanaged forests, particularly where grazing of domestic animals or coppice has been abandoned in recent decades.

Several researchers have highlighted the importance of wood pastures, forest meadows and parklands which, despite obvious anthropogenic influence, contain a rich legacy of ancient trees and forest species of high conservation value (Rose, 1976; Kirby *et al.*, 1995; Vera, 2000; Alexander and Butler, Chapter 17, this volume). There is a potential paradox in the survival of many forest species in managed, semi-open habitats rather than in non-intervention high forest.

In this paper we use palaeoecological data to assess the structure of some temperate forests of north-western Europe during the mid- to late Holocene. We examine this potential paradox in the past, where European pollen diagrams give clear evidence for long-term survival of *Corylus avellana*, *Quercus* and even *Pinus sylvestris*, but with little other evidence for semi-open conditions such as significant presence of herbaceous plants (Bradshaw and Mitchell, 1999; Svenning, 2002). We identify the major agencies that open up forest canopies, and use palaeoecological data to assess their changing importance through time. We adopt the hypothesis that forests will tend to build closed canopies on fertile soils in the absence of disturbance and so we assess the extent and types of past disturbance in north European temperate forest. Studies of the past also help in the development of a 'natural' forest concept by yielding insight into forest conditions at times of reduced anthropogenic impact.

We identify five major groups of disturbance factors that tend to open forest canopies: browsing animals, fire, waterlogging and soils of low fertility, storm and anthropogenic influence.

2.2 Browsing Animals

Vera (2000) highlighted the role of grazing and browsing animals in maintaining open forest conditions in Europe during the Holocene. He has rightly emphasized that many palynologists and forest ecologists have neglected the potential impact of large ungulates on past forest structure, but his specific conclusions have nevertheless generated considerable debate (Svenning, 2002; Bradshaw *et al.*, 2003). One important unknown variable is the size and dynamics of prehistoric ungulate populations, without which knowledge it will be impossible to resolve some of the debated issues. A number of observations, however, strongly indicate that browsing animals alone cannot account for the successful past regeneration and maintenance of significant populations of light-demanding trees such as *Corylus, Quercus* and *Pinus.*

Bison bonasus (bison) and *Bos primigenius* (aurochs) are important ungulates in Vera's discussion, as they have most likely been present in the past in sufficiently large numbers to exert a potential regional impact on forest structure. However, bison did not return to Britain or Ireland during the Holocene (Yalden, 1999), so can be excluded from discussion. Aurochs remains are widespread throughout mainland Britain but have never been found in Ireland, yet regional Holocene vegetation development is very similar on both islands (Fig. 2.1). The Irish Holocene fauna is well studied and the absence of aurochs remains cannot be attributed to an inadequate fossil record (Woodman *et al.,* 1997). The British record suggests populations of a reasonable size, at least in south-east England, that survived until at least 3000 years ago and probably later. The regional pollen diagrams from East Anglia (England) and Antrim (Northern Ireland) are typical of many that could have been selected from these regions. They record a remarkable similarity in the balance between non-forested and forested conditions recorded during the Holocene, and the importance of the light-demanding trees *Corylus, Quercus* and *Pinus. Poaceae* are important at both sites during the late glacial period, when there are many independent indicators of climatically induced open conditions. *Poaceae* are subsequently of minor importance until indicators of anthropogenic activity record forest clearance during the Bronze and Iron ages. *Corylus* and *Quercus* are major trees in both regions, so clearly their Holocene persistence cannot be attributed to aurochs grazing. Svenning (2002) presented comparable evidence from the Danish islands. A *Pinus* phase is recorded in both diagrams, but preceded the increase in *Corylus* and *Quercus* in East Anglia.

This example, taken together with other arguments outlined by Svenning (2002), Odgaard *et al.* (2002) and Bradshaw *et al.* (2003), shows that ungulate pressure alone cannot account for the early to mid-Holocene persistence of light-demanding trees in north-western Europe, although it may well be a contributory factor. We must also examine other disturbance factors that influence forest structure, such as fire.

2.3 Fire

The ecological importance of natural fire is largely recognized in the boreal and Mediterranean zones, but fire has often been regarded as purely of anthropogenic origin within the European temperate forest (Bennett *et al.,* 1990). Several palaeoecological and dendrological studies from southern Scandinavia record the importance of fire in shaping present-day communities, often in

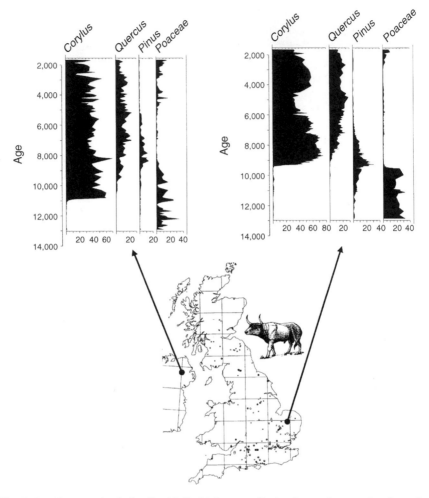

Fig. 2.1. Reported sub-fossil, chiefly Holocene, finds of aurochs remains from the British Isles (after Yalden, 1999) and summary percentage pollen diagrams with a total pollen sum taken from the European Pollen Database, Arles.

conjunction with grazing (Lindbladh *et al.*, 2000; Niklasson and Drakenberg, 2001). We present here plant macrofossil studies from two *Fagus* forest reserves in Halland, south-western Sweden, that show how present forest structure and composition developed after recent cessation of burning and grazing (Hannon, 2002).

Trälhultet today is rather dense forest dominated by *Fagus sylvatica* and planted *Picea abies* with *Alnus glutinosa* in wetter areas. The present structure is dense with a restricted ground flora. Plant macrofossil analysis showed that more open conditions prevailed for at least 700 years prior to the extensive planting of *Picea* during the 1900s (Fig. 2.2). *Picea* macrofossils were first

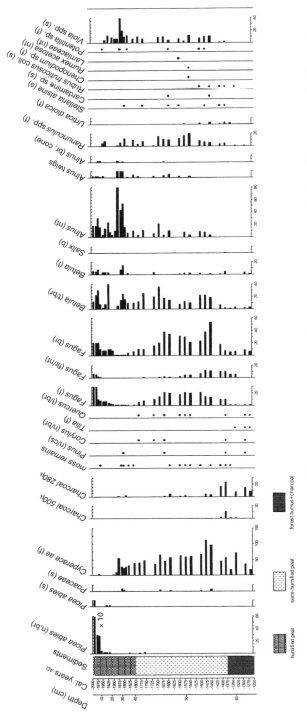

Fig. 2.2. Plant macrofossil remains from Trälhultet, Biskopstorp Nature Reserve, south-western Sweden. Sedimentary details are shown beside the depth profiles and the calibrated age scale. Numbers/50 ml sediment; dots represent single finds; units = 5. Abbreviations: μ, microns; br, bracts; cs, cone scale; flr, flower; f, fruit; l, leaves; n, needle; nt, nut; s, seed.

recorded at the site around 1750 and were probably of spontaneous origin (Hannon, 2002). The ground flora is dominated by wetland taxa growing on the study site, but *Poaceae*, *Rubus fruticosus* coll., *Rumex acetosa* and *Urtica dioica* are typical of taxa that could have grown in open forest on neighbouring drier soils under open conditions. However, it should be noted that *Urtica dioica* and micro-species of *Rubus fruticosus* coll. can also tolerate deep shade. *Corylus*, *Quercus* and *Pinus* were sporadically recorded as macrofossils until around 1850. There are significant amounts of large charcoal fragments recorded at this site between 1250 and 1400, then smaller amounts until 1650 and sporadic occurrence during the early 1800s. These charcoal remains are evidence for fire that contributed to an open forest structure with characteristic vegetation. Fire as the agency for keeping the forest open was replaced by grazing of domestic animals from the 1700s to the mid-1900s and the forest canopy closed over when this practice ceased.

At the Dömestorp *Fagus* reserve, charcoal was recovered from several samples that dated from around 200–300 BC (Fig. 2.3). The only records of *Corylus* and *Pinus* macrofossils at this site came from the same samples. As at Trälhultet, *Poaceae*, *Rubus fruticosus* coll., *Rumex acetosa*, *Urtica dioica* and other indicators of open forest conditions are also found in these samples. Charcoal records virtually cease after this time, but open forest conditions continue, with a rather diverse macrofossil flora. Domestic animals were almost certainly grazing in this forest after the fire period. Their grazing ceased in the recent past and the diversity of herbs declined. *R. fruticosus* coll. increased in importance, as would be anticipated in woods released from trampling and browsing in Britain (Kirby, 2001).

In both these examples, fire and subsequent grazing and browsing had maintained open forest conditions for several centuries in south-western Sweden. Both these disturbance agencies were most likely under close anthropogenic control. However, the importance of fire in south Scandinavian forests can be followed further back through the Holocene at many sites (Lindbladh *et al.*, 2000). Nissatorp in south-eastern Sweden is a typical record, with continuous occurrence of charcoal from 4000 BC until the recent past (Fig. 2.4). The charcoal record is irregular from 4000 BC until about 500 BC, after which time roughly the same number of charcoal fragments per cm^3 was recorded in every sample until the recent past. This charcoal record can be interpreted as an irregular natural fire regime being replaced by more regular, managed burning, associated with slash-and-burn agriculture. This practice ceased, probably during the early 1900s (there is some uncertainty over the dating of recent centuries in the diagram) and *Picea abies* became well established. *Picea* was also extensively planted in this region. The correspondence between charcoal and *Quercus* pollen is striking, and suggests that certain types of fire regime, probably light ground fires, maintained open forest conditions and permitted *Quercus* regeneration (Lindbladh and Bradshaw, 1998).

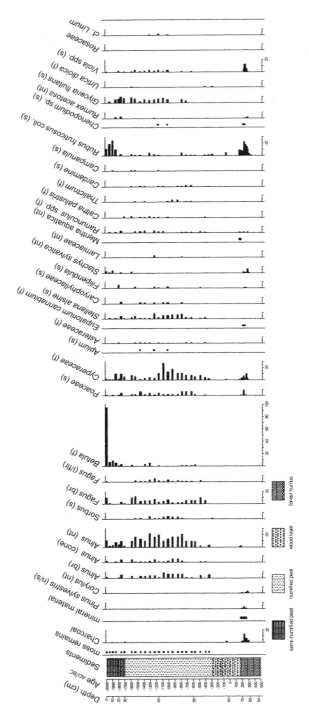

Fig. 2.3. Plant macrofossil remains from Dömestorp *Fagus* reserve. Sedimentary details are shown beside the depth profiles and the calibrated age scale. Numbers/50 ml sediment; dots represent single finds; units = 5. Abbreviations: br, bracts; f, fruit; flr, flower; l, leaves; n, needle; nt, nut; s, seed. Charcoal size range is up to 280 μm.

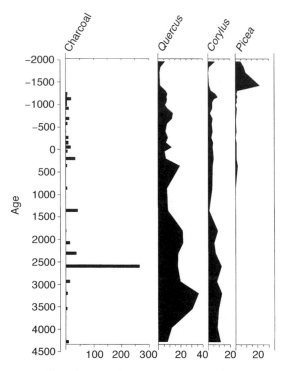

Fig. 2.4. Summary pollen diagram from Nissatorp, south-eastern Sweden. The charcoal data are expressed as fragments per cm^3 and the pollen data as percentages of total pollen.

2.4 Waterlogging, Soils of Low Fertility

While regional water-tables have certainly fluctuated during the Holocene as a result of climate change, there is little doubt that artificial drainage schemes designed to improve agricultural and forest lands have recently altered the hydrological regimes in many temperate forest areas (Rune, 1997). The reduced importance of spring flooding, for example, has almost certainly contributed to a closure of forest structure. It is hard to find conclusive evidence of this effect, but the invasion of Draved Forest, Denmark by *Fagus sylvatica* is a possible example (Wolf, 2003). Studies of long-term succession in Draved show significant recent increases in recruitment of *Fagus* into the forest canopy and a darker, denser forest is being created. *Fagus* grows preferentially on the better-drained soils and on the edges of the drainage ditches. The drainage ditches are now being closed and it will be interesting to observe the consequences for future forest structure. Elsewhere in Europe, recent increases in the importance of *Fagus* have been driven by reduced intensity of management or change from coppice to high forest systems, and these are also possible drivers in Draved Forest.

Sites for pollen and macrofossil analysis are biased in favour of wetter areas and sites, but nevertheless record little flora that are characteristic of open conditions (Odgaard *et al.*, 2002). Suserup Forest, Denmark, is one of the few stand-scale sites where macrofossils have been investigated from the mid-Holocene (Hannon *et al.*, 2000). A rich assemblage of chiefly wetland and fen species was recorded prior to 3200 BC (Fig. 2.5). Its subsequent suppression and replacement by remains of trees and sedges reflects a closing of the forest canopy that may have been partly driven by a lowered water-table or, alternatively, was a natural succession.

Odgaard *et al.* (2002) present a detailed review of the history of the Danish herbaceous flora, characteristic of forest gaps and glades. They emphasize the variation in forest structure associated with soil fertility, showing that herbaceous species are more strongly represented in the pollen record from sandy, well-drained soils with a low clay content. This better herbaceous representation is a strong indication that these sites have more open forest canopies, and their pollen records are not so dominated by arboreal pollen. Sites with thin soil cover on rock outcrops also tend to have naturally open tree canopies. Thus natural variation in soil fertility and drainage are factors that influence forest structure and can generate naturally open, lightly forested areas.

2.5 Storm

Wind-throw of trees is an important agency for generating forest openings, but there is only a limited amount of long-term and palaeoecological data on this topic (Everham and Brokaw, 1996; Webb, 1999; Wolf, 2003). Gap phase replacement of single trees is largely believed to be the dominant dynamic process in European temperate deciduous forest (Emborg *et al.*, 2000) and the typical gaps tend to be subsequently filled with shade-tolerant species. Thus wind-throw has not been promoted as an agency that will enable the long-term survival of *Corylus* and *Quercus* in forests. However, analysis of recent severe storms that affected woodlands where long-term forest monitoring was in progress suggests that storms do also create larger gaps and may have long-term effects on forest structure and composition (Mountford and Peterken, 2000; Wolf, 2003). Data from over 50 years of forest monitoring in Draved Forest, Denmark, showed that wind was the most common factor that removed trees from the canopy (Fig. 2.6). *Alnus* and *Fraxinus* were least affected by wind damage, while well over 50% of canopy removal for *Betula*, *Fagus*, *Quercus* and *Tilia* could be attributed to wind. Wind-throw is clearly another agent that influences forest structure, but, unlike fire and flooding, there is no evidence that it has decreased in importance in recent decades in north-western Europe. We find no reason to attribute major changes in structure and composition to altered wind regime.

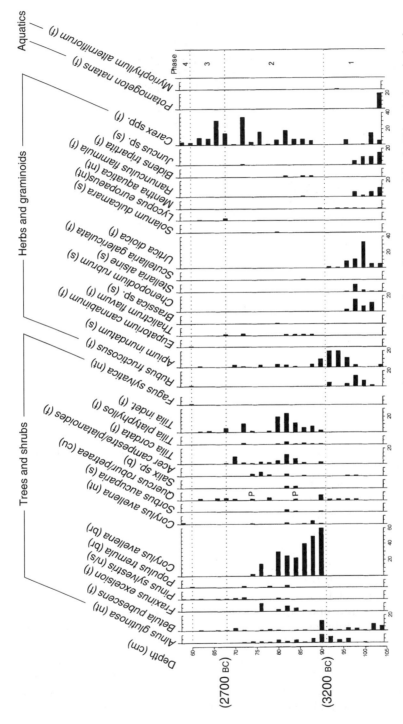

Fig. 2.5. Plant macrofossil remains from Suserup, Denmark. The units are numbers of specimens per 50 ml. Abbreviations: b, bud; br, bracts; cu, cupule; f, fruit; n, needle; nt, nut; s, seed. P beside the histogram for *Quercus* shows where *Q. petraea* macrofossils were found (after Hannon *et al.*, 2000).

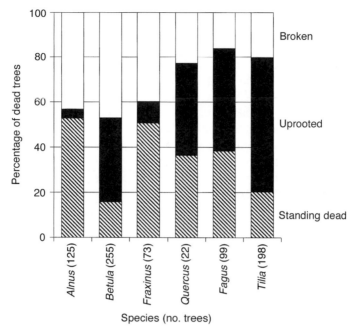

Fig. 2.6.   Types of tree mortality based on a 50-year study of enclosures within Draved Forest, Denmark (after Wolf, 2003).

2.6 Anthropogenic Influence

Many European palaeoecological analyses have shown that anthropogenic impact on forests has been the major influence on forest distribution and composition during recent millennia (Berglund *et al.*, 1996). These studies give few insights into forest structure, and even landscape openness is particularly hard to interpret using regional pollen data alone (Sugita *et al.*, 1999). Selective felling, thinning, burning, litter collection and grazing of domestic stock are the activities most likely to affect forest structure, and all these activities are known to have taken place in temperate European forests.

There has been a widespread and significant reduction in the numbers of domestic grazing animals in European forests in recent decades, although wild ungulates, particularly roe deer, have significantly increased in population size (Fuller and Gill, 2001). These changes have exerted major impacts on forest structure (Bradshaw and Mitchell, 1999). The reduction in domestic animal pressure has certainly contributed to the closure of forests and the loss of open habitat species, but is not a 'natural' process, to the extent that it is under anthropogenic control. Domestic animals can be regarded as having 'replaced' the earlier wild ungulates but, as discussed above, these do not appear to have exerted a widespread control on forest structure. The population sizes of domestic animals were, in certain areas, most likely considerably greater than the former wild populations, but further research is needed to investigate this topic.

Stand-scale palaeoecological investigations are more sensitive to forest structure than regional studies (Bradshaw, 1988), and the increasing number of these sites investigated in north-western Europe is giving new insight into former forest openness. The typical picture to emerge from southern Sweden is of rather closed forest conditions until the first signs of anthropogenic impact. At Mattarp, in south-eastern Sweden, there was a local Viking Age clearance and subsequent abandonment that left a strong signature in the pollen diagram (Fig. 2.7). During the past millennium there is evidence for a variable degree of opening of forest structure, dependent on the intensity of local grazing. Fire was a frequently used tool to improve grazing quality. At the remote Siggaboda, south-eastern Sweden, there was a minimal but observable increase in *Poaceae* pollen and perturbations in the proportion of total tree pollen (Björkman and Bradshaw, 1996) (Fig. 2.7). Anthropogenic influence was greater at the Mattarp site, but has reduced in importance during the past century. We judge anthropogenic influence to have been the dominant driving force for changes in forest structure during the recent past. A trend from extensive to less widespread but more intensive impact in recent years is driving many current changes in forest composition and structure.

2.7 Conclusions

This brief review of the agencies that can affect forest structure and increase openness has identified a number of drivers that have altered through time, with consequent changes in forest structure. Most of the recent changes regarding the reduced importance of grazing animals, the reduced impact of fire and reduced waterlogging of forest soils are a direct consequence of changing anthropogenic practices and impact. It is a difficult task to identify any 'baseline' conditions that preceded significant anthropogenic impact. These conditions altered as climate and forest composition altered, but it seems as if large ungulates only had a minimal impact on forest openness. Natural fires were most likely a significant factor, as were temporary flooding of soils and expression of variations in soil fertility. Combinations of these factors must have created the appropriate conditions for the observed survival of *Quercus*, *Corylus* and *Pinus* in European forests, which were most likely more open and varied than present-day, non-intervention forests. Anthropogenic effects over several millennia have altered these conditions, and a recent trend is forest closure and loss of species that are relatively light-demanding. Conservationists must work with this legacy of millennia of human impact, and the recent focus on wood pasture as a key habitat for the survival of former forest species is a justifiable policy to adopt (Kirby *et al.*, 1995). Wood pasture is unlikely to be the perfect analogue for primeval forest, but it apparently captures some of its

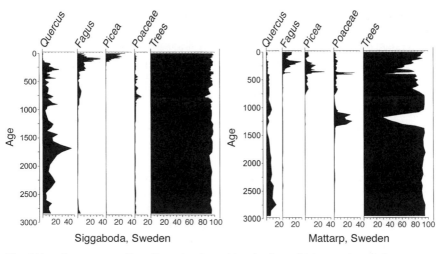

Fig. 2.7. Summary pollen diagrams from Siggaboda and Mattarp in south-eastern Sweden. The data are percentages of total pollen (after Björkman, 1996; Björkman and Bradshaw, 1996).

important characteristics (see also Alexander and Butler, Chapter 17, this volume).

Acknowledgements

We thank The Geological Survey of Denmark and Greenland, Hallands Länsstyrelsen and the Knut and Alice Wallenberg foundation for financial support. Special thanks to Professor Martin Hermy for organizing the conference and commenting on the manuscript.

References

Bennett, K.D., Simonson, W.D. and Peglar, S.M. (1990) Fire and man in post-glacial woodlands of eastern England. *Journal of Archaeological Science* 17, 635–642.

Berglund, B.E., Birks, H.J.B., Ralska-Jasiewiczowa, M. and Wright, H.E. (1996). *Palaeoecological Events During the Last 15,000 Years.* John Wiley & Sons, Chichester, UK.

Björkman, L. (1996) Long-term population dynamics of *Fagus sylvatica* at the northern limits of its distribution in southern Sweden: a palaeoecological study. *The Holocene* 6, 225–234.

Björkman, L. and Bradshaw, R.H.W. (1996) The immigration of *Fagus sylvatica* L. and *Picea abies* (L.) Karst into a natural forest stand in southern Sweden during the last two thousand years. *Journal of Biogeography* 23, 235–244.

Bradshaw, R.H.W. (1988) Spatially-precise studies of forest dynamics. In: Huntley, B. and Webb, T. III (eds) *Handbook of Vegetation Science*, Vol. 7. *Vegetation History*. Kluwer Academic Publisher, Dordrecht, The Netherlands, pp. 725–751.

Bradshaw, R.H.W. and Mitchell, F.J.G. (1999) The palaeoecological approach to reconstructing former grazing-vegetation interactions. *Forest Ecology and Management* 120, 3–12.

Bradshaw, R.H.W., Hannon, G.E. and Lister, A.M. (2003) A long-term perspective on ungulate-vegetation interactions. *Forest Ecology and Management* 181, 267–280.

Emborg, J., Christensen, M. and Heilmann-Clausen, J. (2000) The structure of Suserup Skov, a near-natural temperate deciduous forest in Denmark. *Forest Ecology and Management* 126, 173–189.

Everham, E.M. III and Brokaw, N.V.L. (1996) Forest damage and recovery from catastrophic wind. *The Botanical Review* 62, 113–185.

Fuller, R.J. and Gill, R.M.A. (2001) Ecological impacts of deer in woodland. *Forestry* 74, 189–192.

Hannon, G.E. (2002) *Beech Forest History and Dynamics in Biskopstorp, Halmstads Kommun and Dömestorp, Laholms Kommun, Southern Sweden*. Länsstyrelsen Halland, Halmstad, Sweden.

Hannon, G.E., Bradshaw, R.H.W. and Emborg, J. (2000) 6000 years of forest dynamics in Suserup Skov, a semi-natural Danish woodland. *Global Ecology and Biodiversity Letters* 9, 101–114.

Kirby, K.J. (2001) The impact of deer on the ground flora of British broadleaved woodland. *Forestry* 74, 219–229.

Kirby, K.J., Thomas, R.C., Key, R.S., McLean, I.F.G. and Hodgetts, N. (1995) Pasture-woodland and its conservation in Britain. *Biological Journal of the Linnean Society* 56, 135–153.

Lindbladh, M. and Bradshaw, R.H.W. (1998) The origin of present forest composition and pattern in southern Sweden: a study of the estate where Linnaeus was born. *Journal of Biogeography* 25, 463–477.

Lindbladh, M., Bradshaw, R.H.W. and Holmqvist, B. (2000) Pattern and process in south Swedish forests during the last 3000 years sensed at stand and regional scales. *Journal of Ecology* 88, 113–128.

Mountford, E.P. and Peterken, G.F. (2000) Natural developments at Scords Wood, Toy's Hill, Kent since the Great Storm of October 1987. *English Nature Research Reports* 346, 1–27.

Niklasson, M. and Drakenberg, B. (2001) A 600-year tree-ring fire history from Kvill National Park, southern Sweden – implications for conservation strategies in the hemiboreal. *Biological Conservation* 101, 63–71.

Nilsson, S.G. (1997) Forests in the temperate-boreal transition: natural and man-made features. *Ecological Bulletins* 46, 61–71.

Odgaard, B., Ejrnæs, R., Näsman, U. and Rasmussen, P. (2002) Forudsætningerne for biologisk mangfoldighed i Danmark. In: Møller, P.O., Ejrnæs, R., Höll, A., Krogh, L. and Madsen, J. (eds) *Foranderlige Landskaber*. Syddansk Universitetsforlag, Odense, Denmark, pp. 19–35.

Peterken, G.F. (1996) *Natural Woodland*. Cambridge University Press, Cambridge, UK.

Rose, F. (1976) Lichenological indicators of age and environmental continuity in woodlands. In: Brown, D.H., Hawksworth, D.L. and Bailey, R.M. (eds) *Lichenology: Progress and Problems.* Academic Press, London, pp. 279–307.

Rune, F. (1997) *Decline of Mires in Four Danish State Forests During the 19th and 20th Century.* The Research Series 21, Danish Forest and Landscape Research Institute, Hørsholm, Denmark, pp. 1–93.

Sugita, S., Gaillard, M.-J. and Broström, A. (1999) Landscape openness and pollen records: a simulation approach. *The Holocene* 9, 409–421.

Svenning, J.-C. (2002) A review of natural vegetation openness in north-western Europe. *Biological Conservation* 104, 133–148.

Vera, F.W.M. (2000) *Grazing Ecology and Forest History.* CAB International, Wallingford, UK.

Webb, S.L. (1999) Disturbance by wind in temperate-zone forests. In: Walker, L. (ed.) *Ecosystems of Disturbed Ground.* Elsevier, Amsterdam, pp. 187–222.

Wolf, A. (2003) Tree Dynamics in Draved Forest. PhD thesis, The Royal Veterinary and Agricultural University, Copenhagen, Denmark.

Woodman, P., McCarthy, M. and Monaghan, N. (1997) The Irish Quaternary fauna project. *Quaternary Science Reviews* 16, 129–159.

Yalden, D. (1999) *The History of British Mammals.* Poyser, London.

Landscape Changes in the History of the Austrian Alpine Regions: Ecological Development and the Perception of Human Responsibility

3

E. Johann

University of Agricultural Sciences Vienna, Vienna, Austria

Forest grazing and litter harvesting are examples of close interrelationships between arable land, pastures and forests, tracing back to the first colonization of the mountainous regions. Vegetable manure had a high value for the farmer whose economy was based mainly on stock breeding, and it was as indispensable for the existence of the farm as fuel wood. However, unrestricted practice could not only greatly damage mountain forests but even endanger their existence. This chapter deals with the measures taken by local populations in order to secure the available natural resources on a long-term basis, for the benefit of subsequent generations. In particular, the results of measures aimed at sustainable management have been reviewed and the impact of agroforestry on the preservation of soil fertility, the health of forest stands and the distribution of tree species has been investigated.

3.1 Introduction

Variety in the landscape has always been a natural fact of life for the Austrian population. However, it is often only appreciated if it disappears: if field balks become rarer, if creeks are straightened and hedges are removed. Since 1992, when the 'Convention for Biological Diversity' was negotiated at the United Nations Conference on Environment and Development in Rio de Janeiro, the term 'biodiversity' has been acknowledged internationally. Important aspects of the Convention are, on the one hand, an integrated approach, with the interlinkage of interests in utilization and protection, and, on the other hand, an ecosystem-based approach. It covers all forms of nature, including forests (BMLFUW, 2002; Forum Umweltbildung, 2002).

The biodiversity of an environment is closely related to the local climate and the condition of the soil, including the topography of the landscape. The time scale within which the respective landscape can develop without interference is also of great importance. In the Austrian mountainous areas a great variety of species has evolved in the course of evolution in the richly structured landscape with its high diversity of ecosystems (Kral, 1994; Wilson, 1995; Grabherr *et al.*, 1998).

This chapter emphasizes the historical management of the alpine landscape. Various types of human impacts on all levels of biodiversity are closely interrelated. Societies intervene in natural processes and, in this case, transform them to make them more useful. This kind of interaction with environmental relationships, which has been called 'colonization', is defined as a combination of social activities, modifying certain variables of natural systems and maintaining them in a condition that differs from that of the original form. Thereby natural systems are replaced by others which have a higher output of a specific biomass useful to mankind. Usually periodic or continuous input of labour and expenditure of resources are necessary in order to keep the colonized systems in the desired condition (Fischer-Kowalski *et al.*, 1997).

This chapter highlights the measures that have been taken by the local people in the Austrian Alps in order to secure the available natural resources on a long-term basis for the benefit of subsequent generations. In particular, it reviews the results of efforts to develop sustainable management methods, and investigates the forthcoming future impact of agroforestry on the preservation of soil fertility, the health of forest stands and the distribution of tree species. Previous studies have documented land-use changes and their effects on vegetation patterns (Johann, 1990, 2001, 2003a; Selter, 1995; Schenk, 1996; Bürgi *et al.*, 2000; Schmidt, 2002).

3.2 Background and Approach

For many centuries the forest served other branches of the economy, such as agriculture, manufacturing or hunting (Braudel, 1985). The influence of humans on the forest was dependent on a multitude of different factors and varied locally in relation to process, intensity, time and duration. As the population depended on these resources (Pacher, 1975; Schmidt, 2002), the sustained guaranteed availability of the natural resources was of great concern (Selter, 1995; Radkau, 2000).

In Austria, the period from the Middle Ages to the beginning of the 19th century was characterized by intense forest use, sometimes even turning to heavy exploitation (Johann, 1998). In the history of the Austrian alpine region, human impact on forest composition has taken two main forms: conversion from one type of biological community to another; and modification of the conditions within the biological community. In analysing the ecological development of the Austrian mountain forests, several driving forces have been identified and have been taken into account. The forces can be grouped into four general categories: economics, ecology, politics and management (see Table 3.1). They are all partly interrelated. Some have an impact on the way people managed their immediate environment, whereas others have a more direct impact on the biodiversity of the woodland, in particular those related to the preservation of soil fertility, health of forest stands and the distribution of tree species. Most pronounced was the elimination of broadleaved species, either intentionally or unintentionally, giving rise to the origin of secondary coniferous forests.

Although the large-scale conversion of Austrian mountain forests is widely recognized, little is known about the historical background that has led to the present situation. This chapter analyses human impacts and forest ecosystem responses, taking into account a temporal scale of about 400 years (17th to 20th century).

3.3 Results

3.3.1 Living conditions of the rural population

The living conditions of a rural society are determined, on the one hand, by natural factors such as topography, climate, exposure and altitude and, on the other hand, by social factors, ownership structures and demographic evolution.

Landscape structure

Although the alpine region offers a high variety of locations suitable for the survival of plants and animals, the locations fit for human settlements are very sparse. In the alpine region of Austria particularly, the amount of farmland is limited by topographic conditions. Whereas 18–25% of the total area is unproductive land, the proportion of arable land varies between 1% and 6%, and of meadows between 7% and 21%, of the productive land.

Wheat, rye, barley, oats, beans, potatoes, corn, clover, buckwheat, flax and hay were the main species cultivated until the end of the 19th century. The yield at harvest usually did not exceed three times the amount sown. Growth of most grain was restricted to areas below an altitude of 1600 m above sea level, but barley and rye were restricted to below 1700 m. Meadows offering an annual harvest of hay were restricted to an altitude of 2300 m above sea level.

Table 3.1. Driving forces for ecological changes in the Austrian alpine forests.

Economic	Ecological	Political	Management
Transportation	Forest cover prior to industrialization: changes in population were mirrored by changes in the percentage of land in forest	Ownership	Forest grazing
Industrial activities with potential impact on forests (timber market, iron and salt industry, saw mills)		Population densities	Short-term economic utilization
		Global trade	
			The screening of its economic utility
Increase in human population	Litter harvesting (ground litter and branch litter harvesting)	First and Second World Wars	The requirements of land for food production
Demands for timber, firewood and various wooden products	Deer population		
	Insect diseases		Progress in forest science and implementation of new technologies
Market condition			

At higher altitudes, up to 2450 m, hay could only be harvested once every 2 or 3 years. An altitude of 2600 m must be considered the absolute limit for alpine pastures (Meirer, 1973).

The distribution of crops was influenced by ownership structures, exposure and slope conditions. Usage was determined above all by slope orientation: e.g. on the slopes exposed to the south, grain could be cultivated at a very high altitude, sometimes almost reaching the upper timber line. North-facing slopes were wooded more or less to the bottom of the valleys.

Depending on the local site conditions, a crop rotation economy was practised to a large extent, with rotation periods of 3–15 years (Johann, 2003a). Where arable land was scarce, farming also took place on very steep slopes. This cultivation required a high energy input (e.g. carrying up eroded soil and fertilizer on the farmer's back).

Forest ownership

Forest ownership was highly differentiated and resulted from the colonization processes and from developing social structures. Therefore, large and small properties existed side by side, and the ratio of small and large (more than 100 ha) properties could vary greatly within the Austrian alpine provinces.

Forests were mainly in the hands of farmers (46%), rural and urban communities (26%), the state (16%), and large and small private estates (12%) (Wessely, 1853).

When colonization took place, and the forest was cleared, every farmstead maintained its own forest situated in close proximity – the so-called 'home forest'. This woodland served the needs of the farmstead (firewood, litter, grazing). Initially the farmer was allowed to manage it freely – without restrictions and free of charge. In addition, the communities owned and managed large areas of forests. Community forests were mostly situated just above the settlements or on slopes exposed to the north. Over the centuries, the main bulk of these community forests was divided up. In addition, the farmsteads owned extended rights which were applied in the forests belonging to private and state-owned estates. They were allocated to the farmsteads in order to ensure a sufficient supply of timber, fuel wood, litter and pasture for the farmstead. The term and extent of the respective utilization rights were determined in relation to the size of the farmstead. Due to the lack of grazing land (pastures and alpine pastures) these rights were essential. Often miners and other non-land-owners also had access to forest utilization.

3.3.2 How a high biodiversity arose

Without scientific knowledge, the settlements were established on the most favourable sites. Farmsteads were only built on well-protected sites, and the location of roads was determined by land conditions and exposure of the site. Until the 20th century, most farms were autonomous economical units and only tied to the markets to a limited extent. Because farms were mainly self-sufficient, interconnecting transport routes were less important compared to the present day. Small paths and tracks were quite normal. Creeks were cleared manually, but small woods growing beside these creeks remained untouched because they were able to protect the bank. Because of the high amount of work involved, the straightening of the creeks and streams did not take place. For the same reason, the drainage of meadows was not carried out in many locations. Fields were manually cleared of stones which were deposited at their edges. When the fields were very stony, stone walls arose at the field perimeters. Single trees and shrubs growing in the fields and meadows remained untouched, whether for the purpose of fixing the soil, or as providers of shadow for the livestock and people, or as markers of property boundaries (Hoppichler, 2002).

As self-sufficient units, farmsteads were equipped with woodlands, gardens, fields, meadows and alpine pastures. The spatial distribution of the settlements resulted in a varied landscape, still structurally identifiable to the present day. The utilization of the landscape was adapted to the prevailing conditions and thereby created a high diversity of mountain farmsteads. From the mix of fields and forests, the spatial distribution of the cultivated land, the distribution of meadows and gardens, the establishment of alpine pastures and by forest utilization adapted to the demands of the farmstead, a high biodiversity can be inferred (as an example see Fig. 3.1).

3.3.3 Utilization of the landscape

Peasants' management practices and their effects on the forest condition

In general, the traditional small-scale farming practice, with its emphasis on self-sufficiency, gave rise to a variety of cultural ecosystems. Utilization and adaptation were guided by a variety of cultural plants and animal species, and happened more as a coincidence than as a planned process (Hoppichler, 2002).

   An important part of the rural economy was stock breeding. The diversity of domestic animals with different needs (horses, cattle, cows, goats, sheep, pigs, poultry) belonging to each farm allowed sustainable utilization of the limited resources. The surrounding woodland was indispensable to mountain farming. Farms would not have been able to exist without it. Forests were used everywhere in order to cover the wide variety of requirements of farmsteads, thereby probably inducing a high biodiversity. Grazing in the forest replaced indoor feeding in summer time. Litter harvesting (soil litter and branch litter) contributed to the production of fertilizer. Burning in the forests temporarily increased the amount of arable land. In the summer, woodland below alpine pastures offered the livestock protection against bad weather. The clearing of woodland at the timber line offered the possibility of enlarging the area of alpine pastures. From the 18th century, the sustainable management of

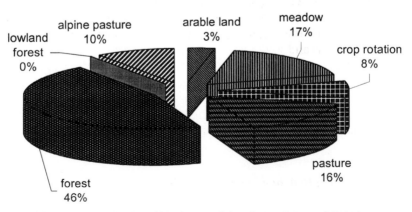

Fig. 3.1. Relative proportion of land uses of the alpine region of Austria (representative example of a small community in the high alpine region; 19th century) (Franzisceischer Kataster Land Kärnten, 1830).

landscape resources was ensured by limiting the number of the cattle that each farm was allowed to drive in the forest, according to the size of the farmstead (Johann, 2003a).

Burning was the favoured method for increasing cropland, and permission to do this was granted to a huge number of mountain farms and acknowledged by Maria Theresia in the 18th century (Johann, 1994). Thereby the farmers had the right of temporary clearance of the woodland for the cultivation of oat and rye, or for grazing. After 2–3 years of cropping, alder, birch, willow and occasionally spruce regenerated naturally. In the middle of the 19th century, burning was still practised on more than 100,000 ha, mostly in Upper Styria, applying to 15% of the forest area and 80% of the farmland (Hafner, 1994).

The most important wooden products were firewood, charcoal, timber, and structural wood for various farm items (such as fences, barrels, wheels). Historical records refer to between 50 and 90 different kinds of wood products (Johann, 2003b). Because wood requirements of the farmsteads varied according to the size of the item and the type of wood needed, selective cutting was the favoured felling method practised by farmers. Therefore, the outcome was an unevenly aged forest, differentiated in its structure and regenerating naturally.

Industrial management practices and their effects on the condition of the forest

The existing salt mines, iron mines, forges and furnaces required a large amount of charcoal, fuel wood and mine props. Since the 16th century regulations had been implemented to ensure a sustainable supply to the mining industry. In this way the amount of wood that was allowed to be cut annually depended on the result of a predetermined inventory. Clear-cutting harvesting methods were legally prescribed in this case, in order to minimize transportation costs. The regrowth of the clear-cut areas depended on natural regeneration. From the beginning of the 19th century onwards, artificial regeneration (seeds and seedlings) was also encouraged.

Clear-cutting methods resulted in the spread of larch (natural), the spread of Norway spruce (by artificial regeneration, mainly from seedlings), the appearance of monocultures of conifers on sites originally dominated by broadleaved plants, and the degradation of forest stands and soils, thereby largely inhibiting the natural regeneration of broadleaves.

Legislation relating to management methods

Soon after early settlement began, rural societies felt the need to regulate the utilization of land interests. As long as ownership and utilization rights were not defined clearly, sustainable management methods and long-term planning could not be implemented in either private or state-owned forests. Therefore numerous laws and regulations were passed by local authorities (farming communities), by the landlords and the government from the 14th century onwards.

When analysing the various laws and regulations passed by the authorities over the centuries, it is clear that most efforts were directed at forest utilization and forest protection: (i) measures on the suitable time of wood harvesting of different tree species; regulations related to burning methods; (ii) regulations concerning forest grazing, in particular of goats, aiming at the protection of the young stands and the protection of specific tree species and cultures. Some tree species, such as oak, areola pine, hazel and also ash and maple, chestnut and even larch, received temporary protection. Special attention was paid to the young stands. Therefore, access to the forest for the purpose of wood cutting and removal was often limited to particular seasons, in order to keep the damage to the young forests to a minimum. Special attention was also paid in local regulations to the utilization of forests that fulfilled protective functions against natural hazards (avalanches, earth slips and flash floods).

3.3.4 Ecological impacts resulting from landscape utilization

The relationship between colonization and sustainability is dialectic. On the one hand, problems of maintaining sustainability presumably arose at the beginning of the development of colonization strategies. On the other hand, colonization processes, in their turn, often put sustainability at risk (Fischer-Kowalski et al., 1997).

As the area of farmland could not be increased by other means, forests became threatened. This was strengthened by the following factors: increasing population, increasing demand for wood products by the mining industry and timber trade, and the nature of the market for agricultural products (the required cash was only secured by extensive stock breeding).

Conflicts arose due to the different demands for utilization on the same forest area. Because the transportation of wood and litter was expensive and also time consuming, those forest stands situated near mining industries and settlements were under the most pressure. One of the long-term problems was the degradation of the soil by soil compaction and nutrient impoverishment, which occurred after intense forest grazing and litter harvesting. Intense grazing in the forest also inhibited regeneration and led to over-thinned forest stands and a reduction of the variety of tree species, in particular broadleaved trees. Forest yield decreased due to the loss of biomass and soil fertility. Frequent pruning caused the decline of timber quality, probably weakened tree health and led to an increase in the number and spread of bark beetle diseases in many parts of the mountainous forests from the beginning of the 19th century. When forests were cleared from steep slopes, soil erosion, landslide, avalanche and flash flood risks followed, thereby threatening not only the farmland, but also the safety of human residents. Transport links (highways, roads and railways) were also frequently damaged by flash floods and gravel cover.

3.3.5 The perception of human responsibility

Relatively little is known about the relationship between farmers and their natural environment. On the one hand, great respect for the power of nature can be observed. Farmers often tried to avert 'evil' or 'misfortune', and to request 'good' fortune, through specific mystic rites. On the other hand, the relationship of farmers to nature can be described as an extremely utilitarian one, within the given framework of social and political limitations. In central Europe, farmers' activities mainly focused on the continued survival of the farmstead. This meant that management methods were generally of low economic risk and that everything was done in order to secure the economic basis of the farmstead (Hoppichler, 2002). Nevertheless, because of their limited resources, farmers were not always able to act in the interests of the longer term.

Not later than the middle of the 19th century, the society collectively became aware of environmental changes caused by excessive utilization and increased industrial development. In Austria, the principles of sustainable forestry were ultimately adopted in practice when social awareness of the link between natural hazards (flash floods and avalanches) and the maintenance of the forest areas increased, not only regionally but also on the national level (Johann, 1995, 2002).

In spite of the recognition that the long-lasting preservation of the living space seemed endangered, there were considerable limitations on the ability of the affected people to act. One of the most important factors was the limited financial resource of the local people. They were not able to introduce changes, e.g. with regard to management methods or with regard to the development of alternative sources of income, although they were perfectly well aware of the close interrelationship between utilization and sustainability. Foresters employed by the government were also familiar with the problem of mismanagement and overutilization, but their remarks were only noticed when a specific Department for Rural Affairs was set up around the middle of the 19th century. However, until the second half of the 19th century, knowledge related to the techniques for the regeneration of wasteland was remarkably poor and scientific investigations had not yet been introduced.

The establishment of a special Department of Agriculture and the transfer of responsibility for forests to this department in 1872 finally led to an improvement in the condition of mountain forests. The most important measures were: the introduction of a forest law that had validity throughout the whole monarchy (1852), the regulation of tenures rights (1853), improved training of the forestry staff, the promotion of scientific investigation, and the introduction of legal and technical measures with regard to the protection of forests and the environment (afforestation of deserted mountain regions and the creation of protection forests) (1884) (Johann, 1994). The active assistance of the local population in afforestation programmes was important, as it achieved maximal acceptance and the active protection of forest stands.

3.3.6 Diversity losses

From the second half of the 19th century onwards, market globalization, especially concerning the import of basic food from the USA, gave rise to market changes that caused changes in land-use management methods.

Changes in land-use management systems

Although the Industrial Revolution had already taken place in the 19th century, traditional farming methods in the alpine areas of Austria remained relatively unchanged even, in spite of liberalism, until the middle of the 20th century. Even though farmers' liberation, diverse crop rotation economies, mechanization and technical improvements (improved seed stock and artificial fertilizer) increased production, the element of self-supply remained. From the middle of the 20th century onwards, industrialization of agriculture introduced mechanization, intensive farming, specialization and rationalization. This resulted in a shift from arable land to pastures, stock breeding in the valleys, indoor feeding throughout the whole year and the abandonment of alpine pastures. As a consequence, unprecedented remarkable structural changes and modifications in the living conditions of the rural population took place, resulting in the abandonment of a considerable number of farmsteads (Johann, 2003a).

The loss of the landscape's biodiversity

The removal of trees and shrubs, the introduction of large-scale machinery, the application of fertilizer, inducing widespread nutrient enrichment, and the use of pesticides led to an escalating loss of the landscape's biodiversity from the 1950s to the 1990s. At present, several government-sponsored programmes encourage agricultural production on an organic basis, with a ban on the use of artificial fertilizers and pesticides. Some mountain farmers are applying these programmes, thereby re-introducing some biological variety once again.

The modification of agriculture resulted in, above all, the enlargement of forested areas, due to the abandonment of farms situated on unfavourable land (steep slopes, alpine pastures). The improvement in the condition of forests was initiated by silvicultural measures, such as an increase in the degree of stocking and the improvement of timber quality when branch litter harvesting came to an end. Afforestation programmes have brought about an alteration in the mix of tree species by the promotion of Norway spruce, often in combination with the setting up of monocultures. Thus the decrease of broadleaved species has continued and larch has begun to regenerate naturally on abandoned pastures (see Fig. 3.2).

3.4 Discussion

At least two major phases can be distinguished in the conversion of European woodland. The first phase was the exploitation until the 19th century, which can be characterized as a 'mining' process carried out by the mining and saw-mill industry. The second phase comprises 19th- and 20th-century forestry, characterized by intensive silviculture and forest management. In order to secure the national timber supply and to protect living space, afforestation of barren land and abandoned farmland was promoted and supported by almost every European country (Bloetzer, 1978; Johann, 2001; Schmidt, 2002). Due to modern forest management systems, securing sustained high interest, monocultures of fast-growing conifers were planted and broadleaved species were eliminated from the 1860s onwards. The rise of secondary Norway spruce forests in Europe mainly goes back to afforestation programmes carried out in the 19th/20th century. The purpose of this kind of forestry was purely economic – nature was not taken into consideration – it had to submit to the 'art of forestry'. In general, the proportion of conifers has increased in almost every European country, with an unintended loss of species diversity. This is

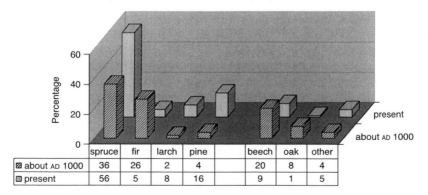

	spruce	fir	larch	pine		beech	oak	other
about AD 1000	36	26	2	4		20	8	4
present	56	5	8	16		9	1	5

Fig. 3.2.   Development of the distribution of tree species in Austria (Kral, 1994).

especially true for the variety of natural tree species and the structure of the stands. With this, the whole scenery changed (Conwentz, 1907).

Whereas in Switzerland, from the turn of the century, the legally prescribed rejection of the clear-cutting economy led to a silviculture based on natural principles, in Austria, as well as in many other European countries, plantation economy continued to be practised in the mountainous regions, apart from some exceptions, during the first decades of the 20th century (Forster, 1954). The loss in the diversity of ecosystems seen today is partly connected to economic–technical developments, the abandonment of agriculture and the increase of forest areas caused by afforestation programmes using one single tree species (mainly Norway spruce). This development is contrary to the wishes of a society seeking a variety of cultural ecosystems and the conservation of open landscapes (Schulz, 1985).

To what extent and to what intensity have human beings influenced the mountain forest ecosystem and how natural are Austrians forests today? The forest area (46.8%) ranks Austria as the second most highly forested country in Europe (BMLFUW, 2002). More than 20% can be classified as semi-natural or natural, according to the very narrow interpretation of Grabherr (1998). Although there is a tendency towards a reduction in the number of tree species, two-thirds of the forest area still have elements characteristic of natural forest. Provinces having a high amount of mountain forest also show the highest portion of natural and semi-natural forests. These provinces are Tyrol and Vorarlberg (inner Alps), Salzburg (northern Alps) and Carinthia (southern limestone Alps). They typically have a historically derived peasant ownership. This remarkably high share of natural and semi-natural forest areas (up to 40%) is only slightly exploited, with limited impact on the ground vegetation and forest structure. These forests have been supported by the development of small-scale forestry, often with characteristic management practices. Controversially, in the valleys and basins of the alpine area, features such as stand stratification or age have been distinctly altered by commercial exploitation and forest grazing. Clear correlations between the degree of alteration and site-related factors such as topography, altitude and gradient have been observed.

The sustainable preservation of the protective (against avalanches and landslides) and welfare functions of forests (including the harmonious balance of the country's water resources and nature conservation) is one of the most important tasks the Austrian mountain forests have to fulfil at the present time (Tersch, 1994). Also, the multiple use of forestry today has to ensure that mountainous regions continue to function as a living space for the rural population, but also to cater for the tourist and recreational activities of the urban society. Only a few years ago there was great concern among the public about the use of nature for recreation purposes and nature protection. Meanwhile, many controversial discussions have arisen about the various sporting activities practised in mountainous areas and the impact of widespread tourism on the landscape in general. In particular, the special type of 'nature tourism', aimed at visiting rare or 'untouched' spots in the landscape, is considered to be in contrast to the goals of nature and landscape protection.

References

Bloetzer, G. (1978) Die Oberaufsicht über die Forstpolizei nach schweizerischem Bundesstaatsrecht. *Zürcher Studien zum Öffentlichen Recht* 2, 222.

Braudel, F. (1985) *Der Alltag. Sozialgeschichte des 15.–18. Jahrhunderts*. München, Germany.

Bundesministerium für Land- und Forstwirtschaft, Umwelt und Wasserwirtschaft (BMLFUW) (2002) *Nachhaltige Waldwirtschaft in Österreich – Österreichischer Waldbericht 2001*. BMLFUW, Vienna.

Bürgi, M., Russel, E. and Motzkin, G. (2000) Effects of post settlement human activities on forest composition in the north-eastern United States: a comparative approach. *Journal of Biogeography* 27, 1123–1138.

Conwentz, H. (1907) *Gesetzliche Vorkehrungen Betreffend den Schutz der Natürlichen Landschaft und die Erhaltung der Naturdenkmäler* 8. Intern. Land- und Forstw. Kongress, Vienna, Referate Sektion VIII–XI, Vol. 4.

Fischer-Kowalski, M., Haberl, H., Hüttler, W., Payer, H., Schandl, H., Winiwarter, V. and Zangerl-Weisz, H. (1997) *Gesellschaftlicher Stoffwechsel und Kolonisierung von Natur. Ein Versuch in Sozialer Ökologie*. Verlag Fakultas Amsterdam, Germany.

Forster, A. (1954) Betrachtungen aus der Praxis zur naturgemäßen Waldwirtschaft im Gebirge. *Allgem. Forstzeitung* 65, 190–191.

Forum Umweltbildung (2002) *Leben in Hülle und Fülle. Vielfältige Wege zur Biodiversität*, Umweltdachverband (Hg.) Vienna, pp. 9–10.

Franzisceischer Kataster Land Kärnten (1830) *Schätzungsdistrikt I, Villacher Kreis*, Kärntner Landesarchiv Klagenfurt.

Grabherr, G., Koch, G., Kirchmeir, H. and Reiter, K. (1998) *Hemerobie Österreichischer Waldöko-Systeme*. Österr. Akademie der Wissenschaften (Hg.), Veröffentlichungen des Österreichischen MaB-Programmes Bd. 17, Universitätsverlag Innsbruck, Austria.

Hafner, F. (1994) Mehrfachnutzung des Waldes. In: Österr. Forstverein (ed.) *Österreichs Wald. Vom Urwald zur Waldwirtschaft*. Vienna, p. 114.

Hoppichler, J. (2002) Leben mit Vielfalt: Landwirtschaft zwischen Intensivierung und Landschaftspflege. Vortragsmanuskript. In: Botanisches Institut der Universität Wien (ed.) *Biodiversität Lernen. Neue Wege zum Schutz der Biologischen Vielfalt*. Tagung Forum Umweltbildung, Vienna, pp. 5–7.

Johann, E. (1990) Forestry as opposed to nature conservation? In: IUFRO (ed.) *Proceedings XIX IUFRO-World-Congress Montreal*. Vienna, pp. 187–198.

Johann, E. (1994) Gesellschaftliche Regelungen zur Walderhaltung und Waldbewirtschaftung. In: Österr. Forstverein (ed.) *Österreichs Wald. Vom Urwald zur Waldwirtschaft*. Vienna, pp. 155–197.

Johann, E. (1995) Naturgewalten. Herausforderung und Antwort in der Geschichte der Österreichischen Forstwirtschaft. *Schweizerische Zeitschrift für Forstwesen* 146/8, 597–609.

Johann, E. (1998) The impact of industry on mountain forests in the history of the Eastern Alps before World War I. In: Diaci, J. (ed.) *Gorski Gozd (Mountain Forest)*. Zbornik Referatov IXI Gozdarski Studenski Dnevi, Univerza v Ljubljani, Slovenia, pp. 81–94.

Johann, E. (2001) Zur Geschichte des Natur- und Landschaftsschutzes in Österreich. Historische 'Ödflächen' und ihre 'Wiederbewaldung'. In: Weigel, N. (ed.) *Faszination Forstgeschichte. Schriftenreihe des Instituts für Sozioökonomik der Forst- und Holzwirtschaft*, Vienna, pp. 41–60.

Johann, E. (2002) Zukunft hat Vergangenheit. In: Österr. Forstverein (ed.) *150 Jahre Österreichischer Forstverein*. Vienna, pp. 158–180.

Johann, E. (2003a) *Waldgeschichte im Nationalpark Hohe Tauern in Kärnten*. Nationalparkdirektion Hohe Tauern Großkirchheim and Kärntner Landesarchiv Klagenfurt, Austria.

Johann, E. (2003b) Die Holzversorgung der Stadt. In: Österr. Akademie der Wissenschaften (ed.) *Umweltgeschichte von Wien*, Kapitel: AG 2.3. Die Versorgung und der Stoffwechsel der Stadt, Vienna.

Kral, F. (1994) Der Wald im Spiegel der Waldgeschichte. In: Österr. Forstverein (ed.) *Österreichs Wald. Vo m Urwald zur Waldwirtschaft*. Österr. Forstverein (Hg.), Vienna, pp. 11–40.

Meirer, K. (1973) Beiträge zur Forstgeschichte Osttirols. Diss. Univ. f. Bodenkultur Wien, Vienna.

Pacher, J. (1975) Entwicklungstendenzen der Forstwirtschaft in Deutschland in der ersten Hälfte des 19. Jahrhunderts unter dem Einfluss allgemeiner Zeitströmungen. *AJFZ* 146, 157–160.

Radkau, J. (2000) *Natur und Macht. Eine Weltgeschichte der Umwelt*. Beck, Munich, Germany.

Schenk, W. (1996) *Waldnutzung, Waldzustand und Regionale Entwicklung in Vorindustrieller Zeit im Mittleren Deutschland*. Steiner Verlag, Stuttgart, Germany.

Schmidt, U.E. (2002) *Der Wald in Deutschland im 18 und 19 Jahrhundert*. Conte Forst Saarbrücken, Germany.

Schulz, W. (1985) Einstellungen zur Natur – eine empirische Untersuchung. Diss. FW. Fakultät München, Germany.

Selter, B. (1995) *Waldnutzung und Ländliche Gesellschaft: Landwirtschaftlicher Nährwald und Neue Holzökonomie im Sauerland des 19. und 19. Jh*. Schöningh Paderborn, Germany.

Tersch, F. (1994) Der Wald in dem wir heute leben. In: Österr. Forstverein (ed.) *Österreichs Wald. Vom Urwald zur Waldwirtschaft*. Vienna, pp. 503–532.

Wessely, J. (1853) *Die Österreichischen Alpenländer und ihre Forste*, Vol. 1. Vienna.

Wilson, E.O. (1995) *Der Wert der Vielfalt. Die Bedrohung des Artenreichtums und das Überleben des Menschen*, Piper Verlag, Munich, Germany.

Natural Tree Colonization of Former Agricultural Lands in the French Massif Central: Impact of Past Land Use on Stand Structure, Soil Characteristics and Understorey Vegetation

4

B. Prévosto,[1] T. Curt,[1] E. Dambrine[2] and P. Coquillard[3]

[1] Cemagref, U.R., Dynamiques et Fonctions des Espaces Ruraux, Campus des Cézeaux, Aubière, France; [2] INRA, Equipe Cycles Biogéochimiques, Seichamps, France; [3] Equipe Gestion de la Biodiversité, Université de Nice, Faculté des Sciences, Parc Valrose, Nice, France

The Chaîne des Puys, a mid-elevation volcanic massif of the French Massif Central, has been subject to large-scale field abandonment in the past five decades, leading to tree colonization, especially by Scots pine and silver birch. Monospecific woodlands have thus established naturally on past croplands, pastures and heathlands. They often exhibit narrow and unimodal distributions of age, and their origin could be explained by a low resistance of the initial ground vegetation after grazing cessation. Woodlands differ in their dendrometric characteristics, floristic composition and soil properties according to past land use. Stands on past croplands are less dense but with a more developed shrub layer, and show a greater abundance of nutrient-demanding species. In these situations the soils have a lower C/N ratio, higher nitrate content and higher nitrate production. Knowledge of former agricultural use appears, therefore, to be a key component in understanding stand development, present vegetation composition and soil fertility of naturally established woodlands.

4.1 Introduction

The French Massif Central has been the subject of a large-scale transformation during the past few decades, like many other mountainous areas throughout Europe (Hüttl et al., 2000). In this region, the collapse of the traditional farming system has led to the abandonment of huge areas that were often colonized by pioneer trees such as silver birch (*Betula pendula* Roth.) and Scots pine (*Pinus sylvestris* L.) which often play a key role in the vegetation dynamics of the mid-elevation mountains of the Massif Central. Their high reproductive capacity, their wide climatic and edaphic range, and their ability to compete successfully with other species in open environments explain why these species are able to colonize an area quickly immediately after a disturbance such as fire or, more frequently, abandonment of agricultural practices. Because of their tendency to form dense thickets, pioneer and invading trees are reputed to cause significant ecological problems for managers of grazing lands and protected areas (Richardson et al., 1994; Barbero et al., 1998). They can cause drastic ecological shifts in the natural or semi-natural vegetation they colonize and deeply alter the nature and functioning of invaded ecosystems (Richardson and Higgins, 1998). Trees can deeply modify vegetation richness and cause local extinction of resident species (Richardson et al., 1996). However, in the Massif Central secondary woodlands provide favourable habitats for wildlife and represent a key step in forest succession towards the ecologically valuable climatic forests, as late-successional species, such as beech (*Fagus sylvatica* L.), naturally establish in these woodlands (Curt et al., 2003). Poorly studied, a better knowledge of these woodlands is therefore needed to better understand their role in forest dynamics and their impact on biodiversity.

Secondary woodlands are specific in that they have formed on lands that were previously not forest but used, under various forms, for agriculture. Past land use has been recognized to have an important impact on soil characteristics (Montes and Christensen, 1979; Tamm, 1991; Hüttl and Schaaf, 1995; Koerner et al., 1997; Dupouey et al., 2002), floristic composition (Peterken and Game, 1984; Hermy, 1994) and stand structure (Foster et al., 1998).

In this study we have analysed how the abandonment of traditional pastoral practices has led to tree colonization in a mid-elevation volcanic mountain range of the French Massif Central: the Chaîne des Puys. More specifically, the impact of past land use on forest structure, soil fertility and ground vegetation was studied for different stands, composed of Scots pine and silver birch, that have established naturally after agricultural abandonment.

4.2 Material and Methods

4.2.1 Study area

The Chaîne des Puys is a mid-elevation volcanic massif composed of a plateau (mean altitude 900 m) dominated by about 100 volcanoes, whose highest point reaches 1465 m (Puy de Dôme). The climate is mountainous with oceanic influences (mean rainfall 1000 mm/year). Volcanic bedrocks are diverse with regard to their physical characteristics, from basalt in blocks to non-consolidated ash-fall deposits, and their chemical composition, which ranges from a high to a low siliceous content. The variability in the bedrock is also reflected in the diversity of the soils. Soils develop through an Andosol formation process, i.e. the formation of non-crystalline compounds (allophanes) and Al/Fe humus complexes through weathering (Shoji *et al.*, 1993). These soils are dark-coloured and characterized by a low bulk density (< 0.7 g/cm^3), a microaggregated structure, a silty texture and the sum (Al + ½ Fe) oxalate-extracted is greater than 2% (Prévosto *et al.*, 2002), which meets the criteria of Andosols (Soil Survey Staff, 1998).

In the first part of the 20th century, woodlands were rare in the Chaîne des Puys. Most of the land was devoted to extensive sheep grazing, whereas some parts were cultivated. But after the Second World War, the traditional agricultural system collapsed due to the intensification and specialization of agricultural practices (Bazin *et al.*, 1983). Figure 4.1 illustrates this afforestation process after the abandonment of traditional livestock grazing in a representative 1100 ha area of the Chaîne des Puys at two dates. In 1954, the landscape was almost completely open. Heathlands and pastures represented more than half of the area, whereas forest cover was weakly developed. By contrast, open areas were severely reduced and natural hazel and birch woodlands were the dominant land use in 1991. As a consequence, colonization by pioneer trees such as silver birch, Scots pine and hazel (*Corylus avellana* L.) occurred on a large scale. Tree colonization has caused a major shift in the landscape of this area. Previous heathlands and pastures have almost

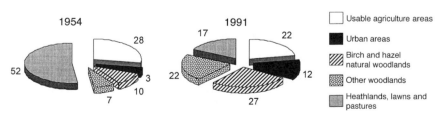

Fig. 4.1. Change of the different land-use categories (in %) in a representative rural district of the Chaîne des Puys (total area 1100 ha) between 1954 and 1991.

disappeared, thus leading to a dramatic change in floristic composition and to the closure of the landscape (Fig. 4.2). Pine stands established preferentially in the southern part of the massif, characterized by a drier climate and a substrate composed of dark-coloured volcanic tephras, whereas birch stands have settled in the central and wetter part of the Chaîne des Puys, on trachytic volcanic tephras (non-coloured tephras with a higher siliceous content).

4.2.2 Data collection

We have selected 27 natural birch and 31 pine stands (mean age ranging from 31 to 45 years) in the central and southern part of the Chaîne des Puys to study their vegetation composition in relation to their past land use. In addition, their soil properties ($n = 12$ and 19) and dendrometric characteristics ($n = 10$ and 15) were also characterized. Past land uses were assessed with the help of the cadastral map of the area established in 1831, completed by historical aerial photographs (1946 and 1954) and an old vegetation map (Lemée, 1959). Three former land-use categories were distinguished: (i) former croplands that were mainly devoted to cereal production and usually composed of small parcels of land surrounded by stone lines; (ii) former pastures that were used mainly for cattle grazing, but which could also have been cultivated temporarily; and (iii) former heathlands that corresponded to larger parcels of land dominated by *Calluna vulgaris* bushes and devoted to extensive grazing. In order to constrain the bias linked to intrinsic differences in soil fertility, woodlands were selected on level areas or gentle slopes at comparable altitude (900–1000 m) and soil depth (40–60 cm). Pine stands are located within a radius of 4 km and birch stands within a radius of 1 km, and mean distance between pine and birch stands is about 11 km.

Vegetation was recorded in 20×20 m plots in which each species was given a coefficient of abundance–dominance according to Braun–Blanquet's procedure (1932). Furthermore, each recorded species was given a coefficient

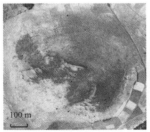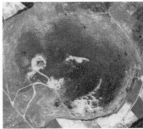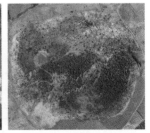

100 m

Fig. 4.2.   Aerial photographs of a volcano colonized by Scots pines in the massif of la Chaîne des Puys at different dates. From left to right, 1962 (before abandonment), 1974 and 1989. Grey areas visible in the two first photographs indicate a ground cover dominated by *C. vulgaris* shrubs.

using Ellenberg's scoring system (Ellenberg *et al.*, 1991, in Lindacher, 1995) that reflected its preferences in terms of soil moisture (F Ellenberg), nutrient availability – especially nitrogen (N Ellenberg), soil reaction (R Ellenberg), light (L Ellenberg), temperature (T Ellenberg) and continentality (K Ellenberg). The average Ellenberg indicator values were then computed for each plot. Syntaxonomical classification of species was also based on Ellenberg *et al.* (1991) in Lindacher (1995).

Soils were sampled at five locations per plot and at two depths (0–15 cm and 15–30 cm), using a single root auger of a fixed volume. Soil subsamples were then mixed (one soil sample/plot/depth), air dried, sieved at 2 mm and analysed for pH, carbon (C), phosphorus (P), nitrogen (N) and nitrate nitrogen (N-NO$_3$). An aerobic incubation of soils was also achieved for the upper layer at constant temperature (15°C) and humidity level (45%) for 6 weeks in order to measure nitrate (NO$_3$) production.

On an area of 400 m^2 per plot, circumferences at breast height of all stems > 1.30 m were recorded and the basal area (m^2/ha) was then computed for each plot. Age at breast height of five dominant trees per plot was determined by tree coring. Moreover, all trees of three additional plots located in three Scots pine stands of increasing age were cored at stem base, and their age determined by tree-ring counting.

4.2.3 Data analysis

Birch and pine stands were analysed separately. The influence of past land use on vegetation composition, soil properties and stand characteristics was then estimated by using a one-way analysis of variance and multiple range test (Tukey's HSD procedure) for each type of woodland (birch/pine). Statistical tests and analyses were performed using the STATGRAPHICS software (v. 5.1 Statistical Graphics Corporation).

4.3 Results and Discussion

4.3.1 Past land use and tree colonization

Most of the natural stands that established after abandonment in this area are monospecific and characterized by a narrow and unimodal distribution of ages (Fig. 4.3), whereas a continuous age distribution, with a large difference between minimum and maximum individual tree age, is a frequent occurrence in naturally regenerated single-species stands (Oliver and Larson, 1996; González-Martínez and Bravo, 2001).

According to a simulation tool we have developed to model the entire afforestation process by Scots pine (Prévosto *et al.*, 2003), narrow age-class

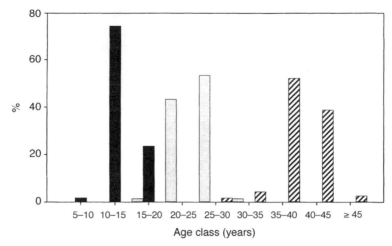

Fig. 4.3. Frequency of age classes in three representative Scots pine stands (respective number of trees sampled: 129, 113, 55) spontaneously established on former pastures.

distributions were produced when trees colonize an area that offered, at least during the first years following its abandonment, favourable conditions for tree settlement, such as an abundant incoming seed rain, a geometry of seed sources that allows seed distribution over the entire area and low resistance by the vegetation in place. In contrast, continuous age distributions were observed, in particular, when the resident vegetation exerted a high and constant resistance to seedling establishment through time. In abandoned pastures and heathlands of the Chaîne des Puys, which are by far the dominant past land uses of this region, the withdrawal of grazing pressure was sudden and massive. Grazing induced a size-reduced vegetation with numerous safe sites for tree settlement. As a consequence, the susceptibility of the ground vegetation to tree colonization was high immediately after abandonment, and woodlands established within a few years (see also Fig. 4.2 for an illustration).

Stands that established naturally on abandoned areas differed in their composition and structure according to past land use. Stands on former croplands were characterized by a more developed shrub-layer of *Corylus* or *Crataegus* shrubs, greater tree dimensions and a reduced tree density, which was half as much in birch stands and a third as much in pine stands as those in woodlands developed in former pastures or heathlands (Table 4.1).

Differences in structure and composition for the same dominant tree species according to past land uses could be related to the differences in soil fertility and vegetation composition (see following sections). Greater nutrient availability observed in former croplands was more favourable to long-lived and perennial herbs that represented a barrier to tree colonization, whereas poorer soils of former heathlands and pastures were more adapted to pioneer

Table 4.1. Main dendrometric characteristics of forest stands in relation to former land use and dominant tree species (birch/pine). Letters indicate statistical differences among the different past land-use categories. Separate analyses were done for birch and pine.

Main tree species Past land use	Silver birch Heathland	Silver birch Cropland	Scots pine Heathland	Scots pine Pasture	Scots pine Cropland
Basal area (m²/ha)	38 (a)	39 (a)	49 (a,b)	55 (b)	38 (a)
Density of the shrub layer (per ha)	200 (a)	2455 (b)	130 (a)	20 (a)	4750 (b)
Density of the tree layer (per ha)	1455 (a)	770 (b)	1630 (a)	1390 (a)	485 (b)
Mean girth (cm)	54 (a)	74 (b)	57 (a)	67 (a)	95 (b)

tree settlement (Rebele, 1992; Smit and Olff, 1998). On the less-fertile soils of abandoned heathlands and pastures, the colonization rate of pioneer and low-nutrient-demanding woody species was higher due to the presence of numerous safe-sites for germination and establishment. Disturbances due to sheep grazing or burning of the vegetation were more frequent in former pastures and heathlands, and such perturbations were recognized as playing a major role in tree invasion (Myster, 1993). In contrast, the more productive soils of past croplands could have favoured the development of a dense sward of perennial species that has hindered the germination and establishment of tree species. These considerations could explain why formerly grazed areas were more quickly and intensively colonized than former croplands after abandonment.

4.3.2 Impact of past land use on soil properties

Past land-use has long been recognized as influencing soil fertility (Montes and Christensen, 1979; Tamm, 1991; Hüttl and Schaaf, 1995; Koerner *et al.*, 1997), but volcanic soils have rarely been studied.

The most visible impact of past land use on soil properties concerned nitrogen dynamics. C/N ratio was inferior in the two first soil layers in past croplands as compared to former heathlands or former pastures (Fig. 4.4A). For the same past land-use, the lower value of the C/N ratio under birch than under pine in the upper layer was due to the species effect: pine litter was in fact more acidifying and resistant to degradation than birch litter, but in the lower soil layer this effect disappeared. Nitrate content was higher in past croplands than in the other past land-use categories (Fig. 4.4B) and nitrates produced by incubation in controlled conditions exhibited the same trend (Fig. 4.4C).

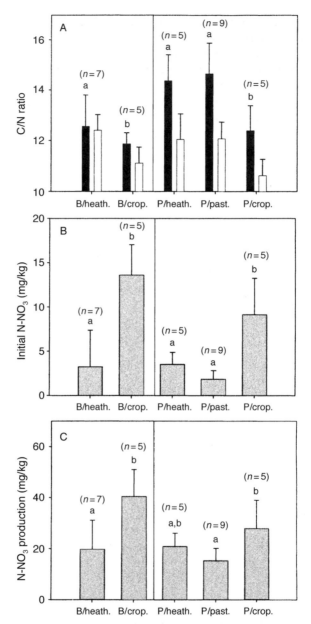

Fig. 4.4.   Soil parameters (mean value and standard deviation) quantifying N supply in relation to the dominant tree species and past land use (*n* = number of samples). The first letter on the horizontal axis refers to the actual tree species (B, birch; P, pine) and the following words to past land use (heath., former heathlands; past., former pastures; crop., former croplands). Letters (a,b) indicate significant statistical differences. (A) C/N ratio of the two upper layers; (B) initial nitrate content; (C) nitrate production after 6 weeks' incubation at 15°C and 45% humidity level.

Former croplands were thus characterized by a higher nitrogen availability, the origin of which lay in past practices, in particular aeration by ploughing and manure inputs. This past land use 'legacy' (cf. Compton and Boone, 2000) was particularly well reflected in our study area by N dynamics, whereas other elements, such as P, were not found to differ significantly among the different past land uses (data not shown). However, at the moment, the exact nature of the mechanisms involved and the time they will persist remain largely unknown.

4.3.3 Impact of past land use on vegetation

We found that species richness under pine or birch was not significantly different between the past land-use categories: the mean number of species per plot of 20×20 m ranged from 45 to 52, according to the past land-use category (mosses were not taken into account). Similarly, Peterken and Game (1984) found that previous land use did not influence the number of species and Hermy (1994) noted that plant species richness was not different between ancient and recent woodlands. In contrast, Koerner et al. (1997) found more species in areas perturbed by human activities than in old forests. This indicates that species richness at the plot level was a fluctuating indicator that did not always respond equally to human perturbations.

However, plant composition differed clearly between the different past land-use categories. Stands that have developed on former heathlands have a significantly higher frequency of species belonging to *Nardo-Callunetea* (Fig. 4.5B), among which the most frequent were *C. vulgaris*, *Deschampsia flexuosa*, *Carex pillulifera* and *Potentilla erecta*. In contrast, former croplands exhibited a higher frequency of species of *Querco-Fagetea* (Fig. 4.5A), such as *Fagus sylvatica*, *Galium odoratum* and *C. avellana*.

Changes in the environment due to past land use were assessed by computing Ellenberg values (Ellenberg et al., 1991). This system has been widely used throughout Europe (e.g. Hill et al., 2000; Dzwonko, 2001) and was used successfully to estimate site modifications due to past human activities (Koerner et al., 1997; Dzwonko, 2001).

Ellenberg mean indicator values for soil moisture, nitrogen and soil reaction were found to be systematically higher in former croplands than in former heathlands (Fig. 4.6). Former pastures have an intermediate position. Ellenberg values for light were only significantly different in pine stands, whereas Ellenberg values for temperature and continentality did not vary with past land use in birch or pine stands (data not shown). These results were coherent with the data collected from soil measurements (see Fig. 4.4) and confirmed that vegetation composition was a synthetic measure of present and past environmental conditions.

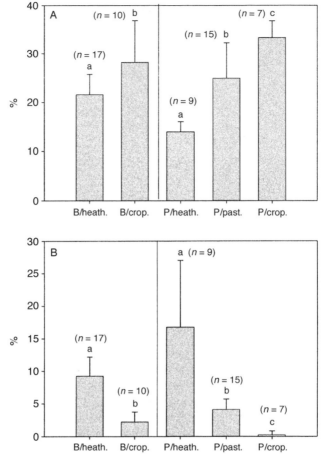

Fig. 4.5. Relative frequency (mean value and standard deviation) of species typical for *Querco-Fagetea* (A) and *Nardo-Callunetea* (B) (Ellenberg *et al.*, 1991) in relation to the dominant tree species and past land use. See Fig. 4.4 for further details.

4.4 Conclusions

The abandonment of the traditional farming system a few decades ago has led to a massive colonization of the studied area by pioneer trees, especially Scots pine and silver birch. The afforestation rate can be explained largely by the sudden abandonment of the grazing regime that represented the main barrier to tree settlement. However, naturally established woodlands differ in their floristic composition, soil properties and stand structure according to past land uses. Woodlands on former pastures and heathlands were quite comparable as regards their dendrometric characteristics and soil characteristics, mainly because their former management was similar (i.e. grazing). However, they

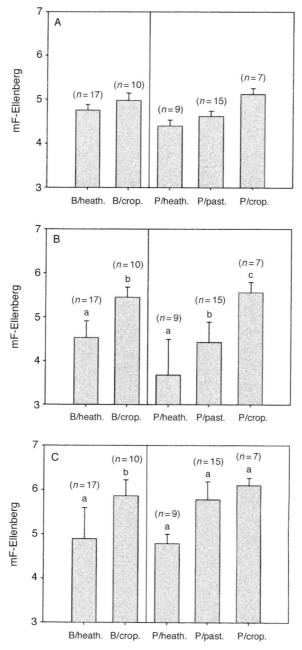

Fig. 4.6. Ellenberg indicator values (mean and standard deviation) in relation to the dominant tree species and past land use. See Fig. 4.4 for further details. (A) Ellenberg value for soil moisture; (B) Ellenberg value for nitrogen; (C) Ellenberg value for soil reaction.

differed with regard to their present floristic composition because the initial resident vegetation also differed. In contrast, woodlands on past croplands exhibited clear differences from those on the other past land-use categories. Most striking differences were a lower tree density, a vegetation in which N-demanding species were more abundant, and higher soil nitrate production and content. Future studies should focus on the vegetation changes in general, and forest species of high ecological value and late-successional tree species (e.g. European beech) in particular.

References

Barbéro, M., Loisel, R., Quezel, P., Richardson, D.M. and Romane, F. (1998) Pines of the Mediterranean Basin. In: Richardson, D.M. (ed.) *Ecology and Biogeography of Pinus.* Cambridge University Press, Cambridge, UK, pp. 153–170.

Bazin, G., Larrère, G.R., de Montard, F.X., Lafarge, M. and Loiseau, P. (1983) *Systèmes Agraires et Pratiques Paysannes dans les Monts Dômes.* INRA, Paris.

Braun-Blanquet, J. (1932) *Plant Sociology. The Study of Plant Communities.* McGraw Hill, New York. (Reprint 1983, Koeltz, Koenigstein.)

Compton, J.E. and Boone, R.D. (2000) Long-term impacts of agriculture on soil carbon and nitrogen in New England forests. *Ecology* 81, 2314–2330.

Curt, T., Prévosto, B., Klesczewski, M. and Lepart, J. (2003) Post-grazing Scots pine colonization of mid-elevation heathlands: population structure, impact on vegetation composition and diversity. *Annals of Forest Science* 60.

Dupouey, J.L., Dambrine, E., Lafitte, J.D. and Moares, M. (2002) Irreversible impact of past land use on forest soils and biodiversity. *Ecology* 83, 2978–2984.

Dzwonko, Z. (2001) Assessment of light and soil conditions in ancient and recent woodlands by Ellenberg indicator values. *Journal of Applied Ecology* 38, 942–951.

Ellenberg, H., Weber, H.E., Düll, R., Wirth, V., Werner, W. and Paulissen, D. (1991) Zeigerwerte von Pflanzen in Mitteleuropa. *Scripta Geobotanica* 18, 1–258.

Foster, D.R. (1992) Land-use history (1730–1990) and vegetation dynamics in central New England, USA. *Journal of Ecology* 80, 753–772.

Foster, D.R., Motzkin, G. and Slater, B. (1998) Land-use history as long-term broad-scale disturbance: regional forest dynamics in Central New England. *Ecosystems* 1, 96–119.

González-Martínez, S.C. and Bravo, F. (2001) Density and population structure of the natural regeneration of Scots pine (*Pinus sylvestris* L.) in the High Ebro Basin (Northern Spain). *Annals of Forest Science* 51, 277–288.

Hermy, M. (1994) Effects of former land use on plant species diversity and pattern in European deciduous woodlands. In: Boyle, T.J.B. and Boyle, C.E.B. (eds) *Biodiversity, Temperate Ecosystems, and Global Change.* NATO ASI series Vol. I. Springer-Verlag, Berlin, Germany, pp. 123–143.

Hill, M.O., Roy, D.B., Mountford, J.O. and Bunce, R.G.H. (2000) Extending Ellenberg's indicator values to a new area: an algorithmic approach. *Journal of Applied Ecology* 37, 3–15.

Hüttl, R.F. and Schaaf, W. (1995) Nutrient supply of forest soils in relation to management and site history. *Plant and Soil* 168–169, 31–41.

Hüttl, R.F., Schneider, B.U. and Farrell, E.P. (2000) Forests of the temperate region: gaps in knowledge and research needs. *Forest Ecology and Management* 132, 83–96.

Koerner, W., Dupouey, J.L., Dambrine, E. and Benoît, M. (1997) Influence of past land use on the vegetation and soils of present day forest in the Vosges mountains, France. *Journal of Ecology* 85, 351–358.

Lemée, G. (1959) *Carte des Groupements Végétaux de la France*. Feuille de Clermont-Ferrand S-O., 1/20,000ème.

Lindacher, R. (1995) *Phanart. Datenbank der Gefässpflanzen Mitteleuropas. Erklärung der Kennzahlen, Aufbau und Inhalt*. H. Sukopp, Zürich, Switzerland.

Montes, R.A. and Christensen, N.L. (1979) Nitrification and succession in the piedmont of North Carolina. *Forest Science* 25, 287–297.

Myster, R.W. (1993) Tree invasion and establishment in old fields at Huchestone Memorial Forest. *The Botanical Review* 59, 251–272.

Oliver, C.D. and Larson, B.C. (1996) *Forest Stand Dynamics*. John Wiley & Sons, New York.

Peterken, G.F. and Game, M. (1984) Historical factors affecting the number and distribution of vascular plant species in the woodlands of central Lincolnshire. *Journal of Ecology* 72, 155–182.

Prévosto, B., Curt, T., Moares C., Dambrine, E., Poutier, F. and Pollier, F. (2002) Les sols sous boisements spontanés de bouleau et de pin sylvestre dans la Chaîne des Puys. Influence du substratum et de l'utilisation ancienne, conséquences sur la végétation. *Etude et Gestion des Sols* 9, 250–267.

Prévosto, B., Hill, D.R.C. and Coquillard P. (2003) Individual-based modelling of *Pinus sylvestris* invasion after grazing abandonment in the French Massif Central. *Plant Ecology* 168, 121–137.

Rebele, F. (1992) Colonization and early succession on anthropogenic soils. *Journal of Vegetation Science* 3, 201–208.

Richardson, D.M. and Higgins, S.I. (1998) Pines as invaders in the southern hemisphere. In: Richardson, D.M. (ed.) *Ecology and Biogeography of Pinus*. Cambridge University Press, Cambridge, pp. 450–473.

Richardson, D.M., Williams, P.A. and Hobbs, R.J. (1994) Pine invasions in the southern hemisphere: determinants of spread and invadability. *Journal of Biogeography* 21, 511–527.

Richardson, D.M., van Wilgen, B.W., Higgins, S.I., Trinder-Smith, T.H., Cowling, R.M. and McKelly, D.H. (1996) Current and future threats to biodiversity on the Cape Peninsula. *Biodiversity and Conservation* 5, 607–647.

Shoji, S., Nanzyo, M. and Dahlgren, R. (1993) *Volcanic Ash Soils*. Elsevier, Amsterdam.

Smit, R. and Olff, H. (1998) Woody species colonisation in relation to habitat productivity. *Plant Ecology* 139, 203–209.

Soil Survey Staff (1998) *Keys to Soil Taxonomy*, 8th edn. USDA-NRCS, USA.

Tamm, C.O. (1991) Nitrogen in terrestrial ecosystems. Questions of productivity, vegetational changes and ecosystem stability. *Ecological Studies* 81, 1–115.

Fire, Death and Disorder in the Forest: 150 Years of Change in Critical Ecological Structures and Processes in Boreal Scandinavia

5

L. Östlund

Department of Forest Vegetation Ecology, Swedish University of Agricultural Sciences, Umeå, Sweden

The history of the boreal forest of Scandinavia is characterized by very low human population density and relatively late impact by modern forestry. The coniferous boreal forest has changed dramatically since the late 19th century, which, in turn, has resulted in loss of biodiversity. Some of the most important ecological changes include fire suppression, loss of dead and dying trees, and loss of multi-storeyed old-growth forest. These changes warrant, among other things, restoration efforts on different spatial scales in order to enhance and protect biodiversity for the future.

5.1 Introduction

Forest history differs markedly between the boreal parts of Europe and the rest of the continent. In the boreal regions, prior to the 19th century, landscapes tended to be dominated by forests and the overall human impact was low (Östlund, 1993). In northern Sweden, land cleared for agriculture covered less than 1% of the total area and human population density was very low (< 1 person/km²) until the late 19th century (Östlund *et al.*, 1997). This contrasts sharply with the deforestation, intensive agriculture and much higher population density across the majority of Europe at the same time (cf. Fritsbøger, 1994; see also many articles in Kirby and Watkins, 1998).

The Industrial Revolution had a major impact on the boreal forests, especially since human impact had been minimal in earlier times. The rapid establishment of a large forestry industry, supplying the rest of Europe with

sawn boards and paper, and the resulting large-scale exploitation of the
old-growth forest in northern Sweden and Finland dramatically transformed
the forest ecosystems and the forest landscape in the late 19th century
(Björklund, 1984; Östlund, 1993). This period of logging the old-growth forest
was replaced, in the 20th century, by the introduction of intensive forest
management. New scientific ideas and new methods were applied, including
large-scale clear-cutting, nitrogen fertilization, draining of wetlands and
the use of herbicides to remove unwanted deciduous trees. These measures
were applied on a broad scale in forest ecosystems that had not, previously,
been subjected to any forest management. This pattern of large-scale logging,
immediately followed by intensive forest management, is comparable to similar processes in North America during the same period (Williams, 1989;
Langston, 1995), but is unlike the events in other parts of Europe.

It is quite clear that these changes in forest use have had a serious impact
on the forest ecosystems at different spatial levels. The forest has changed
dramatically and, in recent times, a number of forest-dwelling species have
been reduced in number or have even become extinct (Berg et al., 1994). We
also know that there is a clear link between changes to the forest habitat and
loss of species.

In this chapter I discuss the large-scale changes that have occurred in the
Swedish boreal forest over the past 150 years. I will focus primarily on such
changes that have affected biodiversity. The results I present are drawn from a
range of sources and are based on either the analysis of biological archives,
such as tree rings, or on historical documents, such as forest inventories and
maps (cf. Östlund and Zackrisson, 2000).

5.2 Threats to Biodiversity in Boreal Sweden Today

At present more than 600 species, primarily cryptogams and invertebrates,
are on the Swedish red-list and are threatened specifically by forestry activities
in the boreal region (Berg et al., 1994). The nature of the threats varies, but
many species are dependent on structures within the forests that have been
severely depleted during the past 100 years (Linder and Östlund, 1998; de
Jong, 2002). Berg et al. (1995) produced a thorough analysis of the factors
underlying the threats to those species currently on Sweden's red-list. A few
major factors were identified (Berg et al., 1995) – the most important being that
modern forestry practices have led to a reduction in the presence of decaying
wood. The second most important factor is that there are fewer old trees, and
the third is habitat destruction. A number of minor factors were also identified,
such as the draining of wet forest land, the closure of the canopy of previously
open forests and a reduction in the number of deciduous trees. While this
assessment helps to provide an understanding of the threats to species and
species groups, its value is somewhat limited because the quality of information

about individual species is variable. In addition, long-term population data are not available for most species. Therefore, such an analysis cannot be very precise, providing only general guidelines about the likely effects of forest management on different species.

It is possible to classify the changes to the forest ecosystems of boreal Sweden. First, there are the direct effects of logging and forest management, such as removal of key elements through clear-cutting of old-growth forest and the removal of dead trees. In addition, there are secondary effects of forest management, such as reduced dead wood production as a result of fire suppression, and a reduction in the number of old trees as a result of shorter rotation periods. These effects, and the factors that affect them, are interrelated, so that it is difficult to explain the decline of a specific niche or element over time. However, it is very important to appreciate the suddenness of the large-scale human impact on the forest ecosystem in this region of Europe. Large-scale logging in the old-growth forests started less than 150 years ago. Forest management began at the beginning of the 20th century, but the more intensive practices have only been in use for 50 years (Östlund, 1993; Östlund et al., 1997).

5.3 Forest Fire: a Key Process in the Boreal Forest

Fire is an essential ecological process in boreal forest ecosystems (Fig. 5.1) (Zackrisson, 1977). Recurrent fires have shaped the ecosystem and maintained vegetation diversity within forest stands and at the landscape level. Furthermore, the native species of these ecosystems have evolved in the presence of frequent forest fires. Numerous studies have documented the historical changes in fire frequency in the boreal forest (Zackrisson, 1977; Engelmark, 1984; Östlund and Lindersson, 1995; Niklasson and Granström, 2000). Zackrisson (1977) showed that until the late 19th century, an average of approximately 1% of the forest burnt annually. In the 1880s, the fire frequency dropped dramatically, because fires were actively suppressed as the timber-frontier advanced across northern Sweden. Such measures were implemented because the timber provided a better economic return than grazing cattle in the forest (Ericsson et al., 2000). In the 20th century, fire was excluded from the ecosystem, except for some prescribed burning in the 1950s and 1960s (Östlund et al., 1997). In the 1990s, FSC-certification, which has been implemented by most of the forest owners in boreal Sweden, has led to an increase in prescribed burning again, this time for conservation purposes.

The consequences of the long period of limited burning in the boreal forest are far-reaching with respect to biodiversity. There have been fundamental changes in vegetation patterns at the landscape level, the spatial heterogeneity has been altered, and certain important niches have been lost. Most fire events in the boreal forest were of low intensity and did not destroy entire stands. The fire frequency at a given site depended on local conditions, exposure and other

Fig. 5.1. Prescribed burning in a multi-storeyed, old-growth forest in northern Sweden. Fire kills some trees, enhances regeneration and is beneficial for a number of threatened plants and animals in the boreal forest. Photo © Lars Östlund.

factors (including anthropogenic influence) and these differences produced a multi-dimensional landscape mosaic (Zackrisson, 1977). Over the larger landscape, a characteristic mosaic of fire-successions was maintained over time. The relationship between the two most important tree-species, Norway spruce (*Picea abies* Karst.) and Scots pine (*Pinus sylvestris* L.), also depended on the fire frequency, since the latter is generally encouraged by fire and the former is generally adversely affected by higher fire frequencies.

In the past, fire had a direct impact on forest stand structure. Fires in pine-dominated forest promoted a multi-storeyed forest, with age classes linked to the frequent fires. Some trees could survive many fires and reached ages exceeding 800 years. Several studies have documented changes in stand structure since the 19th century (Östlund and Lindersson, 1995; Östlund *et al.*,

1997; Axelsson and Östlund, 2001; Axelsson et al., 2002; Hellberg et al., 2003). Östlund et al. (1997), studying vegetation at the landscape level, determined that, in the early 20th century, over 80% of the forest consisted of relatively open, old-growth, multi-storeyed forest. By the 1990s, the cover of this type of forest had been reduced to less than 3%, due to fire suppression and logging activities over less than a century. The new forest taking its place is dense, even-aged and young (Östlund et al., 1997; Ericsson et al., 2000). This transformation of the old-growth forest was the result of a number of different forest management activities. In the late 19th and early 20th century, high-grading of the oldest and largest trees was undertaken. In the early 20th century, thinning and clear-cutting of entire forest stands began, but large-scale clear-cutting was not common until the 1950s. Pre-commercial thinnings, herbicide treatments and the planting of seedlings on clear-cuts were also introduced at this time (Östlund et al., 1997). A general trend throughout the 20th century has been to deal with the 'disorder' of the natural forest and in its place bring order and structure. Another important factor related to fire was induced tree-death. Although stand-replacing fires were uncommon, a number of trees always died even in the low-intensity fires, and therefore a steady supply of dead trees was always available within the ecosystem (Linder and Östlund, 1998).

5.4 Dead Trees and Large Trees: Key Components in the Forest

Prior to the 20th century, the forests of boreal Sweden were 'more grey than green' according to travellers passing through this area. The forest contained so many dead trees that their presence gave a characteristic colour to the forest landscape. The frequent fires killed trees, and trees simply died of old age. Scots pine, the only native pine species in Sweden, can live as long as 800 years, but natural decline and ageing processes usually start after 300–400 years. The corresponding ages for Norway spruce are 400 and 200 years, respectively.

Several studies have shown that there has been a very dramatic decrease in the quantity of dead trees in the boreal forest. One study, covering three different, large, forested landscapes, showed that the quantity of snags was reduced from 11–14 m^3/ha in the late 19th century to less than 1 m^3/ha in the mid 20th century (Linder and Östlund, 1998). Similar decreases have occurred all over the northern Swedish forest and are due to many factors, including fire suppression, exploitation of dead trees for various purposes, 'forest health' issues and the lower overall age of the forest today (Linder and Östlund, 1992). Not only has the actual quantity of dead trees been severely reduced, but also their ecological quality has declined. The current low volume of dead wood in the forest is made up of smaller trees, lacking important ecological qualities associated with larger and older trees. The decay dynamics and wood fungi succession of down-logs is generally size-dependent: smaller snags and down-logs are unable to support a number of threatened species. The loss of habitats associated with dead and dying trees in different forest types and in different habitats is perhaps the most important factor explaining the recent losses in biodiversity (Fig. 5.2) (Berg *et al.*, 1994; Ohlson *et al.*, 1997).

A similar reduction in numbers has occurred with respect to the larger trees. One study has shown that by the late 19th century the density of large trees (diameter at breast height (DBH) > 34 cm) was only 15–20% of its pre-industrial level. Furthermore, really large trees (DBH 50+ cm), which are almost completely absent today, were still quite common in the late 19th century (Fig. 5.3) (Linder and Östlund, 1998). The largest old-growth Scots pines and Norway spruces were the primary targets during the large-scale forest exploitation of the late 19th and early 20th centuries. These trees were old, typically over 200 years, and could reach diameters at breast height of up to 1 m. Their removal across the entire boreal forest resulted in the loss of a key feature, which had characterized the forest landscape. Today's intensive forest management results in forest stands that are clear-cut after approximately 100 years, so large, old trees are never present (Fig. 5.4).

Fig. 5.2. Large snag of a Scots pine, which was killed in a 19th century forest fire. Today's low numbers of snags and down-logs are perhaps the most serious threat to biodiversity in northern Sweden. Photo © Lars Östlund.

Fig. 5.3. Prior to the 20th century, very large and old Scots pines were common in the forests of northern Sweden. Photo courtesy of SCA photoarchive, Sundsvall.

Fig. 5.4. Today's forest, represented by a recently planted clear-cut to the left; and the pre-industrial forest, represented by an old-growth forest in a forest reserve to the right. Photo © Lars Östlund.

5.5 Conclusions

Understanding the full impact of the dramatic changes that have occurred in the boreal Scandinavian forest in recent centuries is not easy. Many changes have taken place at different spatial and temporal scales. Nevertheless, it is important to acknowledge the extent to which the forest has changed, over less than 200 years, from an old-growth natural/semi-natural ecosystem to a managed production forest. It is also important to appreciate that modern forestry practices were implemented over all forest land simultaneously and without exception. Diverse landowners (including the Swedish government, large sawmill and pulp-mill companies, and small, private owners) all employed the same methods. This was the result of very strict forest laws, regulating forest exploitation and forest management, and the emergence of a forestry profession that strongly influenced organizations and companies in the 20th century. It is noteworthy that the transformation of the boreal forest has been extremely successful from a forest manager's point of view. The end result, after almost a century of forest management, is dense, highly productive and economically valuable forests.

There were still very large areas of old-growth forest as late as the early 20th century. Eighty per cent of this could have been described as 'ancient forest', according to criteria applied to forests in nemoral areas of Europe. From the perspective of forest history, such a comparison highlights the great differences between the boreal areas and the rest of Europe.

A number of conclusions, regarding conservation and biodiversity, can be drawn from the above discussion.

1. The initial logging and exploitation of the forests were less detrimental to biodiversity than the intensive forest management in the 20th century. In the earlier phases of exploitation, only the valuable trees were removed, but important ecological structures and processes were left intact in the forest. At the turn of the 20th century, the cut forests may not have displayed the qualities sought by a forester, but they still carried an important biological legacy from the past. The real problems with respect to biodiversity came in the mid 20th century, when intensive manipulation of the ecosystem began. This is a key issue to remember when developing 'New forestry' models.

2. In the boreal forest, very few species are associated with any kind of long-term human use of forest ecosystems, unlike in the nemoral parts of Scandinavia and many other parts of Europe. This does not mean that, over a long period of time, people have not used and influenced the boreal forest in many ways. Indeed, there is evidence of a long history of forest use by the native Samis and other groups in the old-growth forests (Östlund *et al.*, 2002, 2003; Ericsson *et al.*, 2003). However, although widespread, such use was of low intensity and did not create specific ecosystems with specific species constellations. In contrast, the current serious problems related to biodiversity are clearly the result of excessive intensive use of the forests in recent times.

3. It is important, now, that we strive to mimic and restore pre-20th century conditions in the boreal forest. This is essential in order to save and restore biodiversity. Restoration efforts should focus on the processes and structures that have been lost or reduced due to modern forestry practices. I believe that the best tool for restoration is to use historical data about the forests (Nordlind and Östlund, 2003). By understanding how the forest has changed, we can focus on the key issues, identify baseline conditions and understand long-term trends. Restoration will include exciting methods, such as forest burning and killing trees in different ways (Fig. 5.5) (Nordlind, 2001). In addition, certain practices should be avoided by forest managers, including intensive silvicultural methods, such as nitrogen fertilization and planting exotic species.

4. Finally, we now need to focus on broad ecosystem changes and resolve their associated problems, rather than deal individually with every threatened species and its specific requirements. While species-specific research is very important, we simply do not have time to wait for the results.

Fig. 5.5. The killing of a large Scots pine by crushing the cambium with an axe. This can be done with minimal effort when it is very cold in the winter and leaves no axe-marks on the snag. Different restoration methods must be developed and applied in ecologically degraded boreal forest ecosystem in the future. Photo © Lars Östlund.

Acknowledgements

FORMAS, Sweden and the Bank of Sweden Tercentenary Fund supported this work economically.

References

Axelsson, A.-L. and Östlund, L. (2001) Retrospective gap analysis in a Swedish boreal forest landscape using historical data. *Forest Ecology and Management* 147, 109–122.

Axelsson, A.-L., Östlund, L. and Hellberg, E. (2002) Changing deciduous tree distributions in a Swedish boreal landscape, 1820–1999 – implications for restoration strategies. *Landscape Ecology* 17(5), 403–418.

Berg, Å., Ehnström, B., Gustafsson, L., Hallingbäck, T., Jonsell, M. and Weslien (1994) Threatened plant, animal and fungus species in Swedish forests: distributions and habitat associations. *Conservation Biology* 8, 718–731.

Berg, Å., Ehnström, B., Gustafsson, L., Hallingbäck, T., Jonsell, M. and Weslien (1995) Threat levels and threats to red-listed species in Swedish forests. *Conservation Biology* 9, 1629–1633.

Björklund, J. (1984) From the Gulf of Bothnia to the White sea – Swedish direct investments in the sawmill industry of Tsarist Russia. *Scandinavian Economic History Review* 32(1), 17–40.

De Jong, J. (2002) *Populationsförändringar hos Skogslevande Arter i Förhållande till Landskapets Utveckling*. CBM:s skriftserie 7. Swedish University of Agricultural Sciences, Uppsala, Sweden.

Engelmark, O. (1984) Forest fires in the Muddus national park (northern Sweden) during the past 600 years. *Canadian Journal of Botany* 62, 893–898.

Ericsson, S., Östlund, L. and Axelsson, A.-L. (2000) A forest of grazing and logging: deforestation and reforestation history central boreal Swedish landscape. *New Forests* 19(3), 227–240.

Ericsson, S., Östlund, L. and Andersson, R. (2003) Destroying a path to the past – the loss of culturally modified trees along Allmunvägen, in mid-west boreal Sweden. *Silva Fennica* 37(2), 283–298.

Fritsbøger, B. (1994) *Kulturskoven – Dansk Skovbruk fra Oldtid till Nutid*. Gyldendal, Copenhagen, Denmark.

Hellberg, E., Hörnberg, G., Östlund, L. and Zackrisson, O. (2003) Vegetation dynamics and disturbance history in old-growth deciduous forests in boreal Sweden. *Journal of Vegetation Science* 14, 267–276.

Kirby, K.J. and Watkins, C. (1998) *The Ecological History of European Forests*. CAB International, Wallingford, UK.

Langston, N. (1995) *Forest Dreams, Forest Nightmares. The Paradox of Old-growth in the Inland West*. University of Washington Press, Seattle, Washington, DC.

Linder, P. and Östlund, L. (1992) Förändringar i norra Sveriges skogar 1870–1991. *Svensk Botanisk Tidskrift* 86, 199–215.

Linder, P. and Östlund, L. (1998) Structural changes in three mid-boreal Swedish forest landscapes, 1885–1996. *Biological Conservation* 85, 9–19.

Niklasson, M. and Granström, A. (2000) Numbers and size of fires: long-term spatially explicit fire history in a Swedish boreal landscape. *Ecology* 81(6), 1484–1499.

Nordlind, E. (2001) *Restoration of Forests in Boreal Sweden – Gap Analysis and Dead Wood Management at Vitberget, Northern Sweden*. Report of the Department of Forest Vegetation Ecology, Swedish University of Agricultural Sciences, Umeå, Sweden.

Nordlind, E. and Östlund, L. (2003) Retrospective comparative analysis as a tool for ecological restoration: a case study in a Swedish boreal forest. *Forestry* 76(2), 243–251.

Ohlson, M., Söderström, L., Hörnberg, G., Zackrisson, O. and Hermansson, J. (1997) Biodiversity and its lack of correlation with long-term stand continuity in boreal old-growth swamp-forests. *Biological Conservation* 81, 221–231.

Östlund, L. (1993) *Exploitation and Structural Changes in the North Swedish Boreal Forest 1800–1992*. Swedish University of Agricultural Science, Department of Forest Vegetation Ecology, Umeå, Sweden.

Östlund, L. and Lindersson, H. (1995) A dendroecological study of the exploitation and transformation of a boreal forest stand. *Scandinavian Journal of Forestry Research* 10, 56–64.

Östlund, L. and Zackrisson, O. (2000) The history of the boreal forest in Sweden – and the sources to prove it! In: Agnoletti, M. and Anderson, S. (ed.) *Methods and Approaches in Forest History*. CAB International,Wallingford, UK, pp. 34–45.

Östlund, L., Zackrisson, O. and Axelsson, A.-L. (1997) The history and transformation of a Scandinavian boreal forest landscape since the 19th century. *Canadian Journal of Forest Research* 27, 1198–1206.

Östlund, L., Zackrisson, O. and Hörnberg, G. (2002) Trees on the border between nature and culture – culturally modified trees in boreal Scandinavia. *Environmental History* 7(1), 48–68.

Östlund, L., Ericsson, S., Zackrisson, O. and Andersson, R. (2003) Traces of past Saami forest use – an ecological study of culturally modified trees and earlier land-use within a boreal forest reserve. *Scandinavian Journal of Forest Research* 18, 78–89.

Williams, M. (1989) *Americans and Their Forests – a Historical Geography*. Cambridge University Press, Cambridge.

Zackrisson, O. (1977) Influence of forest fires on the north Swedish boreal forest. *Oikos* 29, 22–32.

Relative Importance of Habitat Quality and Forest Continuity for the Floristic Composition of Ancient, Old and Recent Woodland

6

M. Wulf

ZALF e. V. Müncheberg, Müncheberg, Germany

In the Prignitz, which is in the north-west of the federal state of Brandenburg in eastern Germany, ancient, old and recent woodlands show clear differences in site conditions. On average, old woodland occurs frequently on sandy soils, whereas soils in recent woodland are mainly humus-rich, moist or wet, with a high content of nitrogen and basic cations. Nevertheless, the spectra of woodland communities are more or less the same in all woodland types, and range from species-poor communities of the *Quercion* alliance to relatively species-rich communities of the *Alno-Padion* alliance. The discrepancy between information on site conditions from various maps and observed plant communities can easily be explained: maps with a scale of 1 : 25,000 and of 1 : 10,000 cannot demonstrate small-scale, within-forest habitat variation.

Despite comparable spectra of plant communities between all three woodland types, they differed especially in species groups such as typical woodland plants, species with capacity for short- or long-distance dispersal, and species with extremely low and high Ellenberg indicator values. Studies on more or less comparable soils lead to the same trends. Ancient woodland showed a significantly higher share in (extremely) shade- and acid-tolerant species, most of which are exclusively adapted to short-distance dispersal, while recent woods were characterized by light-demanding species. The results from similar soils, especially, supported the assumption that dispersal limitation is a key factor with regard to the association of plants with ancient woodlands.

6.1 Introduction

In general, with respect to forest continuity, woodlands have been divided only into ancient and recent woodland. In Great Britain, the threshold date is the year 1600. Woods that have existed continuously since then are ancient woodland, while recent woodland was established after that point in time (Peterken, 1994). In all other European countries, ancient woodland has been defined as wooded sites for at least 200–250 years. This is due to the lack of maps dating back further than the mid-18th century. Forests established later are all described as recent woodland.

The concept of ancient woodland is well known in various European countries (Brunet, 1993; Dzwonko, 1993; Hermy *et al.*, 1993; Peterken, 1993; Wulf, 1994). Recent woodlands are afforestations on former arable fields, heath- or grasslands; in some cases they were developed by succession on fallow land. Usually, recent woodland has not been divided further into 'older' and 'younger' recent forests. This has neglected the fact that, in many European countries, afforestation practices in the 19th century were directed towards different soil types than in the 20th century. In the 19th century, new woodlands (also called 'old wood') were established predominantly on sandy soil, which had often formerly been used as heath. This was the case in the investigated area and was, moreover, reported by Hermy and Stieperaere (1981) for Belgium, and by Grashof-Bokdam and Geertsema (1998) for The Netherlands. In contrast, more recent afforestations in the course of the 20th century mainly appeared on moist or wet humus-rich soil in lowlands, formerly used as pastures or meadows.

Thus, on the one hand, there is a clear relationship between forest continuity and habitat characteristics, so that the association of some plant species with ancient or recent woodland could be explained predominantly on the basis of site conditions. On the other hand, several studies in ancient and recent woodland, on more or less comparable sites, have revealed differences in species composition (e.g. Van Ruremonde and Kalkhoven, 1991; Heinken, 1998; Bossuyt and Hermy, 2000; Graae, 2000). The potential for short-distance dispersal seems to be the main factor limiting plant species occurrence to ancient woodland. In this context, it has to be remarked, again, that the importance of dispersal capacity depends on habitat quality. The species pool is much higher in soils with a good nutrient and water supply, compared with those in dry, sandy soils. Consequently, due to the larger data set, the significance of correlations between an exclusive short-distance dispersal potential of the species and a high affinity with ancient forests is higher on rich soils than on poor soils (Graae, 2000).

The objective of this study was to assess the relative importance of soil factors and plant life-history traits on the floristic composition of ancient, old and recent forest. More specifically, we addressed the following three questions:

1. Are differences in the floristic composition of ancient, old and recent woodland mainly due to site factors, or could they be better explained with life-history traits of the plant species?

2. Which life-history trait would be most likely to cause differences in the species composition between ancient and recent woodland if the sites were more or less similar?

3. Can differences in the species composition between ancient and recent woodlands on similar sites be observed both on dry, sandy soils and on wet, humus-rich soils?

6.2 Area Investigated

The Prignitz (52°50′N to 53°20′N latitude, 11°30′E to 12°30′E longitude) is situated in the north-west of Brandenburg, one of the federal states in eastern Germany. At the end of the 18th century, about 34% of the landscape was wooded. Towards the end of the 19th century, this had decreased to a minimum of about 18%. Besides deforestations of large wooded areas, some afforestations took place, in most cases on sites less suitable for agriculture, such as heathland and arable fields with a low nutrient content. During the 20th century, woodland area increased to about 23% (nearly 75,000 ha), and new forests were predominantly created on pastures located in valleys. Approximately 85% of all forests are more or less pure Scots pine stands, and only 15% are semi-natural stands. Most of the semi-natural stands are ancient woodland (about 8%), and only 4% and 3% are old and recent woodland, respectively.

6.3 Methods

6.3.1 Data collection

We selected 386 of the semi-natural stands in the study area, of which 223 were ancient, 63 old and 100 recent woodlands. The woodlands are deciduous and mixed woodlands, with an average size of 13 ha. The dominant tree species range from oak–pine, oak–birch and beech–oak stands, with only a few species, on nutrient-poor sites far from groundwater, to oak–hornbeam and alder–ash woods on nutrient-rich sites with groundwater influence.

The continuity of wooded sites was assessed using historical maps dating back to about 1780 (maps of Schmettau; see Lips, 1930/31) and about 1880 (topographical maps of the Prussian government) and actual maps from the 1980s.

Plant species lists were compiled for the 386 selected semi-natural forest communities, ranging from *Quercion* to *Alno-Padion*. The list of plant species includes: (i) all flower and seed plants belonging to deciduous woods and related shrub communities, coniferous forests and related heathland communities or communities of perennial herbs and shrubs growing near woodlands; and (ii) species, which, following Ellenberg *et al.* (1992), show indifferent behaviour or cannot be classified as truly woodland species, although they are only, or mainly, to be found in woodlands in the investigated area (see also Schmidt *et al.*, 2002). In total, 243 species were recorded, of which 30 were trees, 24 were shrubs and 189 were herbaceous species.

The information on the site conditions of the woodlands was compiled from different theme maps. With the help of the Prussian Geological Land Record and other geological maps, the geological substrata of the area were determined for each woodland. Moreover, maps of forest site examinations of the 1950s and 1960s were available. They provided information about site nutritional value and moistness stages, as well as of humus nutritional value and moistness stages.

Moreover, the ecological indicator values L (light), M (moisture), R (soil reaction) and N (nitrogen), after Ellenberg *et al.* (1992), were evaluated based on the species lists, from which conclusions can be drawn for the site conditions with regard to light, water and nutrient conditions. For all mapped plants, details about life form, strategy type and dispersal potentials were assessed using different sources (Müller-Schneider, 1986; Frank and Klotz, 1990; Ellenberg *et al.*, 1992; Frey and Lösch, 1998). To define species with a short- and/or long-distance dispersal potential, the 14 dispersal types after Müller-Schneider (1986) were assigned to three groups of dispersal potential: species exclusively adapted to near- or far-distance dispersal, and species able to disperse over near and far distances (Frey and Lösch, 1998).

A complete overview of all sites showed that sandy soils and humus-rich soils are the most common substrates in the Prignitz. This is in good accordance with the results of the forest inventories. Hence, before two forests sites can be considered as comparable, two selection criteria should be fulfilled: (i) 70% of the area of a wood should be on sandy substratum or on humus-rich wet soils; and (ii) these substrates must absolutely correspond with the site nutritional value and moistness stages, as well as the humus nutritional value and moistness stages. However, these selection criteria could not entirely exclude differences between the woodland types, since the scale of the maps used (1 : 25,000 and 1 : 10,000) was too large to demonstrate small-scale, within-forest habitat variation.

6.3.2 Data analysis

Differences between the three woodland types (abbreviated in tables and figures as ANC, OLD and REC) with regard to their species groups (*Querco-Fagetea* species, species typical and non-typical for woodland), Ellenberg indicator values and life-history traits (life form, strategy type and dispersal potential) were tested with a one-way analysis of variance (ANOVA) with multiple comparisons (Tukey) (Sachs, 1992). The differences of these variables between ancient and recent woodland on comparable sites were tested with the t-test (Zöfel, 1985) with Bonferroni corrections (Cabin and Mitchell, 2000). Differences between the three woodland types with regard to their substrates, site- and humus-types were tested with the Kruskal-Wallis test. Medians of the soil data were compared between groups with Nemenyi tests (Lozán and Kausch, 1998). The association of typical and non-typical woodland species (*Arrhenatheretea-* and *Artemisietea* species) with the three dispersal classes (exclusive short- or long-distance dispersal, and short- *and* long-distance dispersal) was evaluated with χ^2-analysis (Zöfel, 1985). All statistical tests were carried out with the software MINITAB (Anonymous, 2000).

6.4 Results

Although the floristic mapping showed that the entire span of all woodland communities of the Prignitz is present in all three woodland types (Table 6.1), there were clear differences in the species inventory that are closely connected with the site condition (Fig. 6.1a–d).

The most remarkable differences in the flora of ancient, old and recent woodland are caused by species with either extremely low or high indicator values (Fig. 6.2a–d). In the ancient woodland, the high shares in shade-tolerant species (L values 1 to 4) stand out, whereas in the recent woodland there is a large contribution of light-demanding species (L values 7 and 8).

It is obvious that this is due to the occurrence of shade-tolerant true woodland and/or *Querco-Fagetea* species. The ancient woodland plant community is dominated by typical woodland and *Querco-Fagetea* species, whereas in the recent forests, species of *Artemisietea* and *Arrhenatheretea* communities prevail (Fig. 6.3a–c). Many true woodland species are exclusively

Table 6.1. Overview on woodland communities in the Prignitz and their frequencies in the three woodland types ([a]data from Passarge, 1966).

Woodland communities	MSN[a]	NR[a]	(%)[a]	ANC (%)	OLD (%)	REC (%)
Alder–ash–(hornbeam) communities	33.3	12	5.5	13.9	13.1	14.7
Ash–oak communities	25.3	7	3.2	7.7	2.2	6.3
Oak–hornbeam–(beech) communities	17.1	87	39.4	24.1	22.8	21.7
Oak–(pine)–birch communities	17.5	37	9.7	20.7	26.0	14.1
Beech–oak/hornbeam communities	18.4	55	29.8	26.4	28.3	42.6
Beech–(pine) communities	10.2	23	10.4	7.2	7.6	0.7

MSN, mean species number; NR, number of relevés; ANC, ancient woodland; OLD, old woodland; REC, recent woodland.

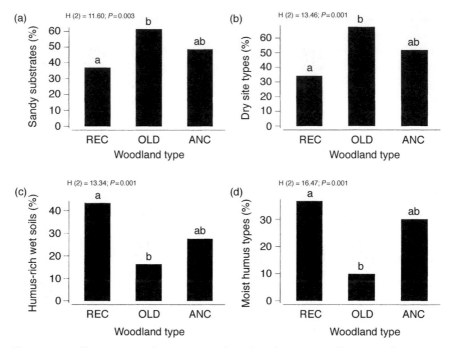

Fig. 6.1. Differences in relative cover of sandy substrates (a), dry site soil types (b), humus-rich wet soils (c) and moist humus soil types (d) between the three woodland types. Medians lacking common letters differ signifcantly at $P < 0.05$.

adapted to short-distance dispersal. Consequently, they are not able to colonize recent forest stands and reach significantly higher contributions in ancient than in old and recent woodland (Fig. 6.3d). In contrast, most of the *Arrhenatheretea* and *Artemsietea* species are good colonizers of recently established forests, able to disperse over long distances, as confirmed by the highly significant result of the χ^2-test (χ^2-value = 42.221, P value < 0.001). Thus, (re)colonization, and also most probably survival on former arable fields and grasslands, of these species are the main reasons for their predominant occurrence in recent woodland.

In general, the differences between ancient and recent woodland on similar soils are more or less the same as demonstrated above. Differences concerned mainly the (very) low indicator values for light, soil reaction and moisture (Table 6.2). However, when the standard Bonferroni correction was applied, significant results were only obtained in species-rich woodlands, although the same trends could be seen for species-poor woodlands on sandy soils (Table 6.2). The data in Table 6.2 show that the number of typical woodland species in ancient woodland is always higher than in recent forests, irrespective of the site qualities.

In most cases the species number is higher in woodlands on humus-rich, moist or wet soils compared to woodlands on dry, sandy soils. Once more,

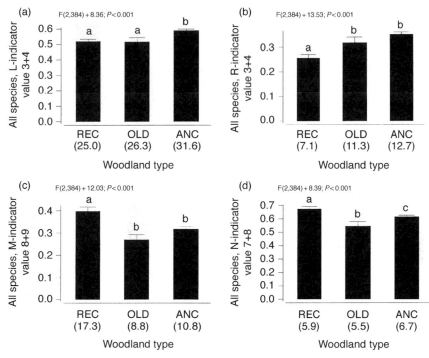

Fig. 6.2. Differences in contribution of herb species with L-indicator values 3 and 4 (a), R-indicator values 3 and 4 (b), L-indicator values 8 and 9 (c) and N-indicator values 7 and 8 (d) between the three woodland types. Values are transformed; means are given in parentheses under the abbreviations of woodland types. Means lacking common letters differ significantly at $P < 0.05$.

one of the most remarkable results is that species with exclusive abilities for short-distance dispersal are clearly more often to be found in ancient woodland than in recent woodland (Table 6.2). Interestingly, the analysis of strategy types doesn't show any significant results, although most true woodland species are less competitive in comparison with species occurring significantly more often in recent woodland (data not shown).

6.5 Discussion

Although we divided recent woodland further into older and very recent woodland, the results are in very good accordance with other studies from different European countries (Dzwonko and Loster, 1992; Dzwonko, 1993; Koerner *et al.*, 1997; Wulf, 1997; Hermy *et al.*, 1999; Bossuyt and Hermy, 2000). Ancient woodland is characterized by, amongst other factors, (very) shade- and acid-tolerant species. The lower indicator values for light of plant species in ancient woodland are correlated with measured light intensities, as

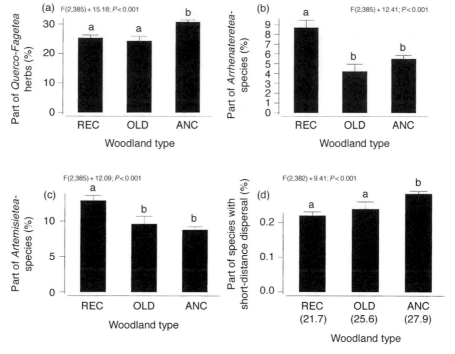

Fig. 6.3. Differences in the contribution of *Querco-Fagetea* herb species (a), *Arrhenatheretea* herb species (b), *Artemisietea* herb species (c) and herb species with exclusive adaptations to short-distance dispersal. Means lacking common letters differ significantly at *P* < 0.05.

has recently been demonstrated by Dzwonko (2001a), and the higher number of species with low indicator values for soil reaction in ancient woodland corresponds very well to measured pH values (Dzwonko, 2001a). Several studies have shown that the soils of ancient woodland are more acidic and therefore support acid-tolerant species (Honnay *et al.*, 1998). In some cases, the higher pH value of soils of recent woodland explains the higher abundance of plant species here.

This was reported for *Mercurialis perennis* L. by Brunet *et al.* (2000). Previous land use, and the associated use of fertilizers, raised the pH values, as well as the content of nitrogen and phosphate (Peterken, 1993; Petersen, 1994; Verheyen *et al.*, 1999). These soil conditions favoured those species non-typical for woodlands, which are much more competitive and may have hampered typical woodland species in their establishment in recent woodland (Dzwonko and Gawronski, 1994; Hermy, 1994; Brunet *et al.*, 2000). This may be of importance to ancient woodland indicators, although it is not suggested by the results from this study.

Nevertheless, this disadvantage of true woodland species is not absolute. Studies on recent woodland adjacent to ancient woodland have shown

Table 6.2. Comparison of indicator values and species groups in ancient and recent woodland on comparable sites (absolute species number, t-test). Due to correction of the significance level by standard Bonferroni only levels < 0.001 are significant; all others would be evaluated as trends.

Variables	Sandy substrates		Humus-rich and wet substrates	
	Means ancient/ recent (n = 62/ n = 24)	Level of significance	Means ancient/ recent (n = 47/ n = 40)	Level of significance
Typical woodland species	11.9/8.2	0.011	15.6/11.0	< 0.001
Typical woodland species with short-distance dispersal	4.4/2.9	0.012	6.0/3.6	< 0.001
Typical woodland species with short- and far-distance dispersal	4.1/2.9	0.031	5.5/4.1	0.007
Typical woodland species with far-distance dispersal	3.4/2.4	0.040	4.1/3.4	0.089
Typical woodland species with				
L-indicator values 1 and 2	1.5/0.8	0.003	1.6/0.8	0.001
L-indicator values 3 and 4	5.9/4.8	0.159	8.5/6.0	0.001
L-indicator values 5 and 6	3.5/1.8	< 0.001	4.0/2.8	0.010
L-indicator values 7 and 8	0.3/0.3	0.964	0.7/0.8	0.526
M-indicator values 4 and 5	7.3/4.3	0.001	7.9/5.0	< 0.001
M-indicator values 6 and 7	3.2/2.8	0.410	5.4/3.9	0.008
M-indicator values 8 and 9	0.6/0.5	0.828	1.3/1.2	0.782
R-indicator values 3 and 4	2.4/1.2	< 0.001	2.2/1.1	< 0.001
R-indicator values 5 and 6	4.3/3.0	0.032	5.6/4.2	0.008
R-indicator values 7 and 8	2.8/2.3	0.296	4.6/3.1	0.013
N-indicator values 2 and 3	1.5/0.8	0.007	1.3/0.8	0.028
N-indicator values 4 and 5	2.9/1.6	0.003	3.1/1.9	0.003
N-indicator value 6	2.5/2.0	0.257	3.2/2.2	0.009
N-indicator values 7 and 8	3.6/2.8	0.175	5.8/4.7	0.027

that, in principle, all species are able to colonize newly established woodland, although, in some cases, the colonization depends on the dispersal mode of plant species and may therefore take several decades (Matlack, 1994; Brunet and von Oheimb, 1998; Bossuyt *et al.*, 1999; Dzwonko, 2001b). Thus, differences in the flora of ancient and recent woodland concerning the dispersal abilities of plants could be demonstrated best in isolated recent forests. As these differences can also be found for ancient and recent woodland on similar sites (see also Dzwonko, 1993), the differences in dispersal spectra are not due to differences in soil characteristics. However, the ability to observe this correlation increased with the size of the species pool, as demonstrated here and by Dzwonko and Loster (1990), Hermy (1992), Dzwonko (1993), and more recently by Graae (2000).

Despite the critical point that dispersal abilities are always derived from the morphology of the diaspores, which may differ from the real dispersal potential, it was demonstrated once more that the low dispersal ability of true woodland species, including ancient woodland indicator species, seems to be the key factor in explaining the association of several plants with ancient woodland. This assumption is supported by several studies on dispersal of diaspores by wild mammals. Irrespective of animal species (hare, martens, roe deer, and wild boar), and dispersal mode (endo- or epizoochorous), those plant species that are non-typical for woodland reach by far the greatest part of the total amount of diaspores in the fur or faeces (Luftensteiner, 1982; Mrotzek *et al.*, 1999; Heinken *et al.*, 2001, 2002; Heinken and Raudnischtka, 2002; Schaumann and Heinken, 2002). Only very few typical woodland species were transported by animals, and the number of diaspores was always extremely low, reaching less than 10% of all diaspores (Mrotzek *et al.*, 1999).

6.6 Conclusions

Differences with regard to the flora between ancient, old and recent woodland caused by site factors primarily concerned species with extremely low or high indicator values for light, moisture, soil reaction and nitrogen. This was also true when ancient and recent woodland on similar soils were compared, whereas on sandy substratum the results could only be evaluated as trends. Besides the importance of site factors, the dispersal modes of plants play an important role in their association with ancient woodland.

Since low dispersal potential seems to be the main factor for no or slow colonization of recent woodland by ancient woodland indicator species, it must be assumed that the exchange of diaspores between forests with different habitat continuity is reduced. Thus, it is unlikely for even just a part of the woodland-typical herb species to form metapopulations. This assumption might apply particularly to the indicator species of ancient woodland. They therefore appear extremely vulnerable to losses in genetic diversity, and it seems to be important to explore further processes such as local extinction, dispersal and colonization against the background of fragmentation and isolation of woodland areas, as well as increasing habitat decline and levelling of site conditions.

Acknowledgements

I would like to thank Christine Daul for her assistance in proof reading and Thilo Heinken and Beatrijs Bossuyt for valuable comments on the manuscript.

Field research was supported partly by the Federal Ministry of Consumer Protection, Food and Agriculture (BMVEL) and by the Ministry of Food, Agriculture and Forestry (MELF) of Brandenburg.

References

Anonymous (2000) *MINITAB*. Release 13 for Windows. Minitab, State College, Pennsylvania, USA.

Bossuyt, B. and Hermy, M. (2000) Restoration of the understorey layer of recent forest bordering ancient forest. *Applied Vegetation Science* 3, 43–50.

Bossuyt, B., Hermy, M. and Deckers, J. (1999) Migration of herbaceous plant species across ancient–recent forest ecotones in central Belgium. *Journal of Ecology* 87, 628–638.

Brunet, J. (1993) Environmental and historical factors limiting the distribution of rare forest grasses in south Sweden. *Forest Ecology and Management* 61, 263–275.

Brunet, J. and von Oheimb, G. (1998) Migration of vascular plants to secondary woodlands in southern Sweden. *Journal of Ecology* 86, 429–438.

Brunet, J., von Oheimb, G. and Diekmann, M. (2000) Factors influencing vegetation gradients across ancient–recent woodland borderlines in southern Sweden. *Journal of Vegetation Science* 11, 515–524.

Cabin, R.J. and Mitchell, R.J. (2000) To Bonferroni or not to Bonferroni: when and how are the questions. *Bulletin of the Ecological Society of America* 81, 246–248.

Dzwonko, Z. (1993) Relations between the floristic composition of isolated young woods and their proximity to ancient woodland. *Journal of Vegetation Science* 4, 693–698.

Dzwonko, Z. (2001a) Assessment of light and soil conditions in ancient and recent woodlands by Ellenberg indicator values. *Journal of Applied Ecology* 38, 942–951.

Dzwonko, Z. (2001b) Migration of vascular plant species to a recent wood adjoining ancient woodland. *Acta Societas Botanicae Polonia* 70, 71–77.

Dzwonko, Z. and Gawronski, S. (1994) The role of woodland fragments, soil types, and dominant species in secondary succession on the western Carpathian foothills. *Vegetatio* 111, 149–160.

Dzwonko, Z. and Loster, S. (1990) Vegetation differentiation and secondary succession on a limestone hill in southern Poland. *Journal of Vegetation Science* 1, 615–622.

Dzwonko, Z. and Loster, S. (1992) Species richness and seed dispersal to secondary wood in southern Poland. *Journal of Biogeography* 19, 195–204.

Ellenberg, H., Weber, H.E., Düll, R., Wirth, V., Werner, W. and Paulißen, D. (1992) *Zeigerwerte von Pflanzen in Mitteleuropa*, 2nd edn. Goltz, Göttingen, Germany.

Frank, D. and Klotz, S. (eds) (1990) *Biologisch-ökologische Daten zur Flora der DDR*, 2nd edn. Martin-Luther-Univ. Halle-Wittenberg, Wissenschaftliche Beiträge 1990/32, Germany, p. 41.

Frey, W. and Lösch, R. (1998) *Lehrbuch der Geobotanik*. Gustav Fischer, Stuttgart, Germany.

Graae, B.J. (2000) The effect of landscape fragmentation and forest continuity on forest floor species in two regions of Denmark. *Journal of Vegetation Science* 11, 881–892.

Grashof-Bokdam, C.J. and Geertsema, W. (1998) The effect of isolation and history on colonization patterns of plant species in secondary woodland. *Journal of Biogeography* 25, 837–846.

Heinken, T. (1998) Zum Einfluss des Alters von Waldstandorten auf die Vegetation in bodensauren Laubwäldern des Niedersächsischen Tieflandes. *Archiv für Naturschutz und Landschaftsforschung* 37, 201–232.

Heinken, T. and Raudnitschka, D. (2002) Do wild ungulates contribute to the dispersal of vascular plants in central European forests by epizoochory? A case study in NE Germany. *Forstwissenschaftliches Centralblatt* 121, 179–194.

Heinken, T., Hanspach, H. and Schaumann, F. (2001) Welche Rolle spielt die endozoochore Ausbreitung von Pflanzen durch Wildtiere in Wäldern? Untersuchungen an Wildschwein (*Sus scrofa*), Reh (*Capreolus capreolus*), Damhirsch (*Cervus dama*) und Feldhase (*Lepus europaeus*) in zwei brandenburgischen Waldgebieten. *Hercynia* N. F. 34, 237–259.

Heinken, T., Hanspach, H., Raudnitschka, D. and Schaumann, F. (2002) Dispersal of vascular plants by four species of wild mammals in a deciduous forest in NE Germany. *Phytocoenologia* 32(4), 627–643.

Hermy, M. (1992) Compositional development of deciduous forests from non-forest precursors in northern Belgium: evidence from historical ecology. In: Teller, A., Mathy, P. and Jeffers, J.N.R. (eds) *Responses of Forest Ecosystems to Environmental Changes*. Elsevier Applied Sciences, London, pp. 437–444.

Hermy, M. (1994) Effects of former land use on plant species diversity and pattern in European deciduous woodlands. In: Boyle, T.J.B. and Boyle, C.E.B. (eds) *Biodiversity, Temperate Ecosystems, and Global Change*. Springer, Berlin, pp. 123–144.

Hermy, M. and Stieperaere, H. (1981) An indirect gradient analysis of the ecological relationship between ancient and recent riverine woodlands to the south of Breges (Flanders, Belgium). *Vegetatio* 44, 43–49.

Hermy, M., van den Bremt, P. and Tack, G. (1993) Effects of site history and woodland vegetation. In: Broekmeyer, M.E.A., Vos, W. and Koop, H. (eds) *European Forest Reserves*. Pudoc Scientific Publishers, Wageningen, The Netherlands, pp. 219–232.

Hermy, M., Honnay, O., Firbank, L., Grashof-Bokdam, C. and Lawesson, J.E. (1999) An ecological comparison between ancient and other forest plant species of Europe, and the implications for forest conservation. *Biological Conservation* 91, 9–22.

Honnay, O., Degroote, B. and Hermy, M. (1998) Ancient-forest plant species in western Belgium: a species list and possible ecological mechanisms. *Belgium Journal of Botany* 130, 139–154.

Koerner, W., Dupouey, J.L., Dambrine, E. and Benoit, M. (1997) Influence of past land use on the vegetation and soils of present day forest in the Vosges mountains, France. *Journal of Ecology* 85, 351–358.

Lips, K. (1930/31) Zur Entstehung der Schmettauschen Karte des Preußischen Staates. *Mitteilungen des Reichsamt für Landesaufnahme* 6, 208–210.

Lozán, J.L. and Kausch, H. (1998) *Angewandte Statistik für Naturwissenschaftler*, 2nd edn. Parey Buchverlag, Berlin.

Luftensteiner, H.W. (1982) *Untersuchungen zur Verbreitungsbiologie von Pflanzengemeinschaften an vier Standorten in Niederösterreich*. E. Schweitzerbart'sche Verlagsbuchhandlung, Stuttgart, Germany.

Matlack, G.R. (1994) Plant species migration in a mixed-history forest landscape in eastern north America. *Ecology 75*, 1491–1502.

Mrotzek, R., Halder, M. and Schmidt, W. (1999) Die Bedeutung von Wildschweinen für die Diasporenausbreitung von Phanerogamen. *Verhandlungen der Gesellschaft für Ökologie* 29, 437–443.

Müller-Schneider, P. (1986) *Verbreitungsbiologie der Blütenpflanzen Graubündens*. Veröffentlichungen des Geobotanischen Institutes ETH, Stiftung Rübel (Zürich) 85.

Passarge, H. (1966) Waldgesellschaften der Prignitz. *Archiv für Forstwesen* 15, 475–504.

Peterken, G.F. (1993) *Woodland Conservation and Management*, 2nd edn. Chapman and Hall, London.

Peterken, G.F. (1994) The definition, evaluation and management of ancient woods in Great Britain. *NNA-Berichte* 3/94, 102–114.

Petersen, P.M. (1994) Flora, vegetation, and soil in broadleaved ancient and planted woodland, and shrub on Rosnaes, Denmark. *Nordic Journal of Botany* 14, 693–709.

Sachs, L. (1992) *Angewandte Statistik*, 7th edn. Springer, Berlin.

Schaumann, F. and Heinken, T. (2002) Endozoochorous seed dispersal by martens (*Martes foina*, *M. martes*) in two woodland habitats. *Flora* 197, 370–378.

Schmidt, M., von Oheimb, G., Kriebitzsch, W.-U. and Ellenberg, H. (2002) Liste der im norddeutschen Tiefland typischen Waldpflanzen. *Mitteilungen BFH* 206, 1–37.

Van Ruremonde, R.H.A. and Kalkhoven, J.T.R. (1991) Effects of woodlot isolation on the dispersion of plants with fleshy fruits. *Journal of Vegetation Science* 2, 377–384.

Verheyen, K., Bossuyt, B., Hermy, M. and Tack, G. (1999) The land use history (1278–1990) of a mixed hardwood forest in western Belgium and its relationship with chemical soil characteristics. *Journal of Biogeography* 26, 1115–1128.

Wulf, M. (1994) Überblick zur Bedeutung des Alters von Lebensgemeinschaften, dargestellt am Beispiel 'historisch alter Wälder'. *NNA-Berichte* 3/94, 3–14.

Wulf, M. (1997) Plant species as ancient woodland indicators in north-western Germany. *Journal of Vegetation Science* 8, 635–642.

Zöfel, P. (1985) *Statistik in der Praxis*. UTB Gustav Fischer, Stuttgart, Germany.

Land-use History and Forest Herb Diversity in Tompkins County, New York, USA

7

K.M. Flinn and P.L. Marks

Department of Ecology and Evolutionary Biology, Corson Hall, Cornell University, Ithaca, New York, USA

A legacy of past agriculture continues to shape many current forests in north-eastern North America and Europe, where the understorey flora of postagricultural stands shows reduced diversity and altered composition for hundreds of years. Effective management of forests recovering from human disturbance requires an understanding of the mechanisms that produce these patterns. Here we discuss forest regeneration on abandoned agricultural land in Tompkins County, New York, USA, in comparison with other regions that share a broadly similar history. Using data from vegetation surveys of primary and secondary forests, we assessed the relationship between land-use history and forest herb diversity; the effects of time since abandonment, distance from primary forest and type of agricultural use, whether cultivation or pasture, on the species-richness of secondary stands; and the influence of dispersal mechanism on individual species' distributions. Secondary forests 70–100 years old contained, on average, 65% the number of forest herb species in primary forests. Among all secondary stands, forest herb species-richness increased with time since abandonment, and decreased with distance from primary forest; type of agricultural use did not affect species-richness directly, but the relationship with distance was weaker for former pastures. Dispersal mechanism did not influence species' relative frequency in the two forest types. While landscape-level patterns indicate that the process of dispersal largely controls forest herb diversity in postagricultural stands, unexplained variation among species with similar dispersal mechanisms suggests that other processes, such as recruitment, may also affect the distributions of some species.

7.1 Introduction

Many present landscapes of north-eastern North America and Europe have been shaped by a history of agriculture (Whitney, 1994). Current plant communities in these regions show the lasting impact of analogous patterns of past land use, in which phases of forest clearance were followed by field abandonment and forest regeneration. As stands develop after agriculture, the recovery of the species-richness of shade-tolerant perennial herbs is of particular interest because the several hundred species that grow exclusively in forest habitats represent the majority of plant diversity in these communities. Life-history traits typical of these species, such as short seed dormancy, low seedling establishment rates, slow individual growth and long pre-reproductive periods, may also make them vulnerable to habitat fragmentation (Bierzychudek, 1982). In landscapes where human disturbance transformed much of the forest, understanding the processes that control the recovery of species richness is critical to the preservation of diversity in remaining woods and its restoration in regenerating stands.

A substantial body of work from both north-eastern North America and Europe now documents that the understorey flora of these postagricultural, or secondary, forests remains less species-rich than that of forests that were never cleared, or primary forests, for hundreds of years (e.g. Peterken and Game, 1984; Dzwonko and Loster, 1989; Matlack, 1994; Singleton *et al.*, 2001). The forest herb communities of primary and secondary forests also continue to differ markedly in composition, as some species characteristic of primary forests frequently occur in secondary stands, whereas others remain absent (e.g. Whitney and Foster, 1988; Wulf, 1997; Brunet and von Oheimb, 1998; Bossuyt *et al.*, 1999). Two main hypotheses have been proposed to explain the observed patterns of plant colonization of secondary forests: dispersal limitation, in which forest herb species' ability to disperse to new sites controls their distributions, and recruitment limitation, in which species' ability to establish and persist in new habitats controls their distributions (Honnay *et al.*, 2002). These mechanisms may operate simultaneously, and they probably interact; while the establishment of forest herb species may largely depend on the arrival of seeds and spores, for instance, the open conditions and dense herbaceous vegetation of old fields probably prevent establishment for several decades after abandonment. Both dispersal and recruitment limitation also take place at multiple scales, so that patterns of variation among landscapes in the extent of their recovery, among forest stands in their species richness and composition, and among individual species in their colonization capacity all help to discriminate among the alternative hypotheses.

Predictions that landscape characteristics, such as the extent and timing of forest clearance and regeneration, should determine the amount and rate of recovery imply that dispersal limitation drives forest herb species distributions (Vellend, 2003). At the same time, variation among landscapes in the relative quality of forest habitats may also contribute to differences in the degree to which they recover. Within regions, analogous expectations apply to the scales of forest communities and individual species. If the occurrence of species is limited primarily by dispersal, then attributes of stands such as area, isolation from potential source populations and time since abandonment should predict their species richness, and dispersal distances should discriminate among species in their frequency. If, on the other hand, occurrence is limited by recruitment, then the quality of sites should predict their richness, and the frequency of species should follow their habitat requirements and demographic rates.

Here, to test these hypotheses, we present work from Tompkins County, New York, USA as a case study in forest regeneration on abandoned agricultural land, and we compare it with other regions that share a broadly similar history.

7.2 Location: Land-use History

Tompkins County, located in the Finger Lakes region of central New York, USA, and approximately 35×35 km (Fig. 7.1), was sparsely inhabited by Iroquois in the 1700s. According to original land-survey records, 99.7% of the county's land area was forested in 1790, primarily with beech, maple and basswood (Marks and Gardescu, 1992). Widespread clearance for agriculture began with European settlement soon after the survey, continued throughout the 1800s, and peaked in 1900, when agricultural census data show that forest cover reached a minimum of 19% (Smith *et al.*, 1993). As farmers abandoned agricultural land throughout the 1900s, the area in forest increased to 51% of the county by 1980 (Smith *et al.*, 1993), and since then forest cover has continued to increase. Over half of the county's forests are thus secondary and, of those, over half originated within the past 60 years.

This short, simple and well-documented history of changes in forest cover makes Tompkins County ideal for the study of the effects of past land use on current forest plant communities. Little forest clearance predates the earliest records; Native American populations apparently had little impact on forests here, relative to other parts of North America that were more densely settled or influenced by fires (Day, 1953). Thus sweeping changes in land use have taken place only within the past 200 years, and detailed records are available from several key points during this time. Mature forests recognizable on 1938 aerial photographs (Fig. 7.2a), the earliest for the county, would have been present at the time of maximum clearance; these sites have almost certainly remained forested throughout the historical period. The first and subsequent photographs, taken in 1954, 1968, 1980 and 1991 (Fig. 7.2b), also show secondary forest stands in various stages of development on old fields, allowing estimation of the time of abandonment and measurement of the distance to the nearest primary forest. Site history and former use as arable or pasture can be further

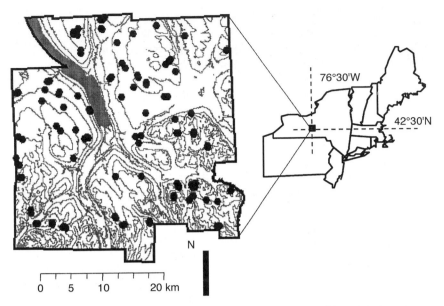

Fig. 7.1. Map of Tompkins County and its location in central New York and north-eastern USA. Contour lines at 60-m intervals show the range in topography from the lake in the north-west, at 120 m elevation, to the hills in the south, at 600 m. Dots mark 111 of the forest sites sampled in the studies synthesized here; six additional sites that were included in analyses could not be relocated precisely.

established by interviews with landowners and field observations of evidence such as stone walls, barbed wire and treefall pits and mounds (Marks and Gardescu, 2001).

7.3 General Comparisons: Tompkins County and Other Landscapes

Tompkins County differs from other landscapes in several attributes that we expect to affect the recovery of forest herb diversity in predictable ways. First, more primary forest remains in Tompkins County than in most of lowland Europe (e.g. Peterken and Game, 1984; Dzwonko and Loster, 1989), more developed parts of north-eastern North America (Nyland *et al.*, 1986; Matlack, 1997), and more fertile parts of the midwest where agriculture remains productive (Curtis, 1956; Whitney and Somerlot, 1985). The region also has more agricultural land that was abandoned to forest regeneration than in more developed and more fertile areas, but not as much as in places such as central Massachusetts, USA, where abandonment was nearly complete (Foster, 1992). Relative to most other cultural landscapes, then, Tompkins County has high proportions of its land area in both primary and secondary forests. The

Fig. 7.2. Aerial photographs from 1938 (a) and 1980 (b) of a 2 × 1.5 km area in south-western Tompkins County, New York, USA. The 1938 photograph shows several rectangular patches of primary forest in the centre; adjacent to these, abandoned farm fields with scattered young trees, and fields in active agriculture: the light-coloured ones arable, others probably pasture. Substantial hedgerows surround many fields. By 1980, secondary forest covers much of the area, and fields continue to fill with trees. Dark-coloured conifers stand out against leafless deciduous trees in this photograph, taken in April.

time of clearance and abandonment differed from other regions as well; land was both cleared and abandoned later here than in Europe and much of north-eastern North America (Williams, 1989). As the recent time-frame of land-use change in Tompkins County has allowed little time for forest recovery, stand age may largely control the extent to which secondary forests have regained species richness.

The spatial configuration of forest habitat also affects the ability of under-storey plants to migrate in the landscape. Tompkins County and other parts of north-eastern North America tend to have much connectivity among habitat patches, as secondary stands often expand and connect primary forests. In northern Tompkins County, for instance, 61% of the stands and 89% of the area of secondary forest that originated before 1938 were adjacent to primary forest, and much of the southern part of the county is covered by contiguous areas of forest as large as several km^2 (Smith *et al.*, 1993). Even in 1900, when forests on the flat, rich land of northern Tompkins County were most frag-mented, random points in farm fields were only typically 100–300 m from the nearest primary forest (Smith *et al.*, 1993). Fields not close or directly adjacent to primary forest may nevertheless be connected to older woods by the network of hedgerows that often surrounds fields in the region. Rather than the fence lines or planted hedges common in Europe, most hedgerows in Tompkins County and north-eastern North America more generally represent remnants of primary forest, left in swathes typically 5–7 m wide when the land was cleared on either side. These often support relict populations of forest herbs that may spread into adjacent fields (Corbit *et al.*, 1999). Given the proximity of forest herb populations in both primary forests and hedgerows, stand isolation may be less important in regulating the species richness of secondary forests in Tompkins County than in landscapes that remain more fragmented.

Additional factors that influence the recovery of forest diversity in a landscape are the interval of time between the peak of fragmentation and the beginning of regeneration and the intensity of agricultural use the land experienced during that interval. In Tompkins County, the interruption in otherwise continuous forest cover was relatively brief. Much of the oldest secondary forest grows on parcels of land that were among the last to be cleared in the late 1800s and the first to be abandoned in the early 1900s, as they were never well-suited to cultivation. Particularly in the southern part of the county, which has more hilly terrain and more acidic soils than the northern part, many farms were marginally productive at best (Neeley, 1965). Since agriculture was unprofitable on these lands, inputs of manure and other improvements may have been low as well. After 70–100 years, in fact, cultivated and pastured soils have recovered sufficiently in their physical and chemical properties to be indistinguishable from adjacent soils that were continuously forested (K.M. Flinn, M. Vellend and P.L. Marks, unpublished data). This pattern contrasts with those found in places where the legacy of agriculture on soils and vegetation has lasted hundreds or thousands of years (Compton and Boone, 2000; Dupouey *et al.*, 2002), and it indicates that the imprint of past land use in Tompkins County may be light *vis-à-vis* regions that supported agriculture of greater duration and intensity.

Whether a parcel of land was pastured or cultivated, different types of agricultural use also have unique and lasting effects on soils and species composition. As dairying predominates in Tompkins County and much of north-eastern North America, agricultural lands include both pasture, often unploughed, and arable, where most field crops are cattle fodder: maize, hay and oats (Neeley, 1965). Ploughing changes the physical properties of soils by mixing soil horizons, destroying pits and mounds, and promoting erosion, while inputs of fertilizer and exports of crops alter nutrient concentrations. Although grazing animals churn or compact soils by trampling and the deposition of manure, in general they disturb soils much less than growing crops. Former pastures may resemble forests sooner than former cultivated fields in canopy cover as well, where wolf trees remain or woody vegetation invades before grazing ceases (Marks, 2001; Marks and Gardescu, 2001).

Relict populations of forest herbs also persist in some pastures that were never ploughed, so that the understorey flora may recover more quickly after pasturage than after cultivation (Rackham, 1975; Stover and Marks, 1998).

Finally, forest diversity depends on the quality of the available habitats, which are often modified by use as woodlots. In this regard, both primary and secondary forests in Tompkins County, and much of north-eastern North America, have been less intensively managed than many forests elsewhere. Nearly all of the county's primary forests were selectively cut and grazed by cattle, but they experienced much less human disturbance than ancient woodlands in Europe that were planted and coppiced as well (Rackham, 1980). Also in contrast to many recent woodlands in Europe, regeneration of secondary forests in Tompkins County was largely natural; the percentage of the county's forest area that was planted, mostly conifer stands started in the 1930s, ranged from only 1% in the north to 9% in the south (Smith *et al.*, 1993). While planting speeds forest regeneration in general, some commonly planted conifers, such as Norway spruce, cast deep shade that inhibits understorey growth.

To summarize, we expect forest communities in Tompkins County to recover more fully than those in regions with less primary forest, less connectivity and longer and more intense agricultural use – indeed, as fully as those in any region that continues to support a cultural landscape of human habitation and active agriculture. Only 100 years have passed, however, since the time of maximum clearance; it is unclear how long full recovery of forest diversity may take.

7.4 Specific Predictions: Forest Herb Diversity within Tompkins County

1. Species richness of forest herbs in secondary forests should increase with time since abandonment, and this relationship should be strong due to the short time since regeneration began.
2. Species richness should decrease with distance from primary forest, but this relationship may be weak due to the high connectivity of habitat patches in the landscape.
3. Former pastures should contain more forest herbs than former cultivated fields, due to relict populations in sites that were never ploughed.
4. Frequency of occurrence in secondary forests should vary with dispersal mechanism, with species carried by vertebrates and wind more frequent than those carried by ants and gravity.

The dependence of species richness on time, distance or type of agricultural use will support the hypothesis that dispersal limitation controls the species richness of secondary forests. If dispersal mechanism predicts frequency of occurrence, this will indicate that dispersal is also a dominant process in determining individual species distributions; remaining variation in colonization ability will suggest species-specific recruitment limitation.

7.5 Methods

We synthesized data from 25 years of vegetation surveys throughout Tompkins County (Stover and Marks, 1998; Singleton *et al.*, 2001; S. Gardescu, unpublished data; P.L. Marks and S. Gardescu, unpublished data; C.L. Mohler and P.L. Marks, unpublished data; M. Vellend, unpublished data). Sites sampled included 47 primary forests and 60 secondary stands that ranged in age from one to 100 years since abandonment, where the total area sampled per site ranged from 180 to 1000 m². The analysis focused on shade-tolerant perennial herbs characteristic of local mesophytic forests (Corbit, 1995), including only the 50 taxa of vascular plants identified consistently in all studies. *Circaea*, *Galium* and *Viola*, as well as all grasses and sedges, were excluded because some studies omitted or failed to distinguish these taxa, and spring ephemerals were excluded because some sites were sampled in late summer after they had senesced. In order to control for differences among studies in the area of quadrat samples and the total area sampled per site, we calculated sample-based rarefaction curves for all sites using *EstimateS* (Colwell, 2000; Gotelli and Colwell, 2001). These were then rescaled by area to compare species richness at a common area sampled, 180 m². We compared the species richness of primary and secondary forests using ANOVA, and with multiple regression we assessed the effects of time since abandonment, distance from primary forest and type of agricultural use, whether cultivation or pasture, on the species richness of secondary stands. Another factor in the multiple regression took into account one additional difference among sampling regimes of separate studies: the area quadrat samples were spread across. Species richness was natural-log transformed, and one outlier was removed (as in Stover and Marks, 1998). Using ANOVA, we assessed the effect of dispersal mechanism on species' relative frequency of occurrence in primary and secondary forest sites, grouping vectors of seed and spore dispersal into four classes: ants; wind; gravity, including explosion; and vertebrates, whether by adhesion or ingestion (Matlack, 1994; Corbit *et al.*, 1999; Singleton *et al.*, 2001). These analyses were conducted with general linear models in the GLM procedure of SAS (SAS Institute, Cary, North Carolina, USA). Finally, we analysed individual species' patterns of occurrence in primary and secondary forests with G-tests of independence corrected for multiple comparisons at $\alpha = 0.05$ by the sequential Bonferroni method.

7.6 Results and Discussion

Although the species richness of the understorey flora in secondary forests 70–100 years old remained lower than in primary forests in Tompkins County

(F = 21.4, df = 1, P < 0.0001; N = 80 sites), the recovery of diversity has been substantial. These secondary forests, the oldest in the region, contained on average 65% the number of forest herb species in primary forests (Fig. 7.3).

Time since abandonment, distance from primary forest and type of agricultural use, including one significant interaction and controlling for area sampled across, explained 77% of the variation in species richness of secondary stands (overall F = 36.1, df = 5, P < 0.0001; N = 60 sites). Species richness clearly increased with time since abandonment (Fig. 7.4a), which accounted for the greatest share of variation by far (F = 96.1, df = 1, P < 0.0001). Both former cultivated fields and former pastures remained quite poor in forest herb species until 30–40 years after abandonment; this time may represent a threshold when old fields generally develop a woody thicket with a closed canopy and thus begin to resemble forests in the habitat they provide for understorey plants. Secondary stands then appear to gain steadily in forest herb species richness, and 100 years after abandonment the process does not seem to have reached an equilibrium.

The relationship between distance from primary forest and species richness was less clear (Fig. 7.4b). The number of forest herb species in secondary stands did decrease with distance (F = 6.1, df = 1, P = 0.016), but this relationship was stronger for former cultivated fields than for former pastures (F = 5.5, df = 1, P = 0.022). Although type of agricultural use *per se* did not affect forest herb species richness (F = 2.2, df = 1, P = 0.15), the history of sites in cultivation or pasture thus affected the colonization process indirectly. Among former cultivated fields, stands adjacent to primary forests showed the full range of variation in species richness, whereas isolated stands had similarly

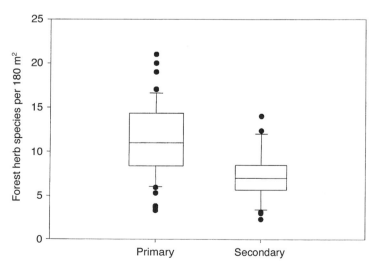

Fig. 7.3.   Species richness of forest herbs in primary forests (N = 47) and secondary forests 70–100 years old (N = 33) in Tompkins County, New York, USA.

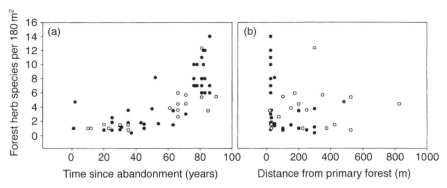

Fig. 7.4. Effects of time since abandonment (a) and distance from primary forest (b) on species richness of forest herbs in secondary stands in Tompkins County, New York, USA. The figure shows scatterplots of raw data without other effects factored out. Closed circles represent former cultivated fields (N = 42), open circles former pastures (N = 19). One former pasture site, marked with a star, was removed from statistical analysis as an outlier (as in Stover and Marks, 1998).

low species richness whether they were at great or small distances from primary forests. This apparent difference between adjacent and isolated forests may be more important than any differences among sites in degree of isolation, particularly in a landscape where few sites were more than 300 m from primary forest. For former pastures, on the other hand, whether sites are adjacent or isolated may be less important than whether sites themselves contain relict populations of forest herbs. One former pasture, excluded from statistical analysis as an outlier, illustrated this pattern particularly well. Although 475 m from primary forest, the site had never been ploughed, and healthy populations of 13 forest herbs were present after over 50 years in active pasture and only 2 years of fallow (Stover and Marks, 1998). The chance that relicts survive forest clearing and subsequent grazing may depend on the specific history and microtopography of patches within sites, such as their proximity to wolf trees and bedrock outcrops (Bellemare *et al.*, 2002). If the colonization of former pastures by forest herbs is a mostly local process, it may still take time for relict populations to spread throughout a site, but distance from more remote sources may have little effect. The occurrence of forest herb populations in hedgerows surrounding both types of abandoned lands may also weaken the effect of distance from primary forest on the species richness of secondary stands (Corbit *et al.*, 1999).

Dispersal mechanism did not affect species' relative frequency of occurrence in primary and secondary forests ($F = 1.87$, df = 3, $P = 0.15$; $N = 38$ species). Several possible explanations may account for this lack of association between dispersal mechanism and colonization ability. First, the traditional classes of dispersal vectors may inadequately represent species' true ability to migrate in the landscape (Vellend *et al.*, 2003). Secondly, the preponderance of stands adjacent to primary forests and the abundance of relict populations in pastures and hedgerows may allow species with short-distance dispersal mechanisms to spread into secondary forests as often as those with long dispersal distances. Finally, species with similar dispersal mechanisms varied in their ability to colonize secondary forests (Table 7.1). Among ferns, for instance, with ostensibly equivalent capacities for long-distance spore dispersal, some species, such as *Polystichum acrostichoides*, occurred much more frequently in primary forests, while others, such as *Athyrium filix-femina*, occurred with comparable frequency in the two forest types (Table 7.1). Dispersal limitation fails to explain this pattern; in fact, *P. acrostichoides* populations in primary forests deposited tens of thousands of spores per m^2 in adjacent secondary stands over a single growing season (K.M. Flinn, unpublished data). Instead, these fern species may differ in their ability to complete subsequent life history stages, such as spore germination or sporophyte establishment, in secondary forest habitats. Such cases suggest that the legacy of past land use may also include persistent environmental changes that inhibit the germination, establishment, growth or reproduction of some species.

The effects of time and distance, factors that control the recovery of diversity at the levels of whole landscapes and forest stands – and that here explain much of the variation in species richness – are now quite well documented (e.g. Peterken and Game, 1984; Matlack, 1994; Honnay *et al.*, 2002; Vellend, 2003). In Tompkins County, stand age was the dominant predictor of forest herb diversity, suggesting that time alone may restore species richness to abandoned agricultural lands without further human intervention. Broad-scale patterns of species richness in this landscape and elsewhere are thus consistent with the hypothesis of dispersal limitation. However, much of the variation among individual species remains unexplained. The dispersal mechanism of forest herb species in Tompkins County did not predict their ability to colonize secondary stands, yet species showed striking differences in their relative frequency in the two forest types. The process of recruitment limitation, though more subtle, may be critical for certain species, so that active restoration may be necessary for secondary stands to gain the species composition of the understorey community in primary forests. Correlations between life history traits and colonization ability begin to address this issue (Verheyen *et al.*, 2003), but many interesting ecological questions about particular species' interactions with land-use history remain open, and work on these should complement landscape-level perspectives in a full understanding of the process of forest regeneration after agriculture.

Acknowledgements

Much gratitude is due to Sana Gardescu, whose good work undergirded all aspects of this review; to Mark Vellend, whose consistent insight clarified many

Table 7.1.   Percentage frequency of occurrence of forest herbs in primary and 70–100-year-old secondary forests in Tompkins County, New York, USA. The table includes the 38 taxa with ≥ 5 occurrences. Bold type indicates taxa significantly more frequent in primary forest, according to G-tests of independence corrected for multiple comparisons at α = 0.05 by the sequential Bonferroni method.

Taxon	Dispersal mechanism	Primary (%) ($N = 47$ sites)	Secondary (%) ($N = 35$ sites)	G	P
Prenanthes spp.	**Wind**	**55**	**9**	**21.45**	**< 0.0001**
Polystichum acrostichoides	**Wind**	**74**	**29**	**17.61**	**< 0.0001**
Aster divaricatus	**Wind**	**81**	**40**	**14.68**	**0.0001**
Trillium grandiflorum	**Ants**	**70**	**31**	**12.41**	**0.0005**
Actaea alba/rubra	**Ingestion**	**53**	**17**	**11.71**	**0.0007**
Hepatica acutiloba	**Ants**	**28**	**3**	**10.44**	**0.0015**
Thalictrum dioicum	Gravity	26	3	9.22	0.0028
Thelypteris noveboracensis	Wind	40	11	9.02	0.0030
Solidago caesia	Wind	40	14	7.01	0.0088
Dryopteris marginalis	Wind	21	3	6.91	0.0099
Polygonatum pubescens	Ingestion	53	26	6.41	0.0121
Sanguinaria canadensis	Ants	30	9	5.98	0.0158
Uvularia sessilifolia	Ants	60	34	5.20	0.0237
Oxalis montana	Explosion	9	26	4.44	0.0381
Aralia nudicaulis	Ingestion	30	11	4.18	0.0434
Caulophyllum thalictroides	Ingestion	38	20	3.26	0.0736
Dryopteris carthusiana/ intermedia	Wind	74	57	2.71	0.0790
Ranunculus abortivus	Gravity	17	6	2.59	0.1145
Trillium erectum	Ants	32	17	2.37	0.1280
Podophyllum peltatum	Ingestion	72	57	2.05	0.1560
Asarum canadense	Ants	11	3	1.99	0.1707
Hydrophyllum virginianum	Gravity	19	9	1.89	0.1766
Geranium maculatum	Explosion	38	26	1.46	0.2320
Athyrium filix-femina	Wind	28	17	1.28	0.2655
Osmorhiza claytonii/longistylis	Adhesion	13	6	1.20	0.2867
Aster acuminatus	Wind	9	3	1.22	0.2869
Smilacina racemosa	Ingestion	70	60	0.93	0.3402
Maianthemum canadense	Ingestion	66	57	0.66	0.4203
Trientalis borealis	Gravity	23	31	0.65	0.4238
Allium tricoccum	Gravity	17	11	0.51	0.4814
Medeola virginiana	Ingestion	23	17	0.49	0.4917
Aster macrophyllus	Wind	13	9	0.37	0.5520
Uvularia perfoliata	Ants	13	9	0.37	0.5520
Waldsteinia fragaroides	Gravity	13	9	0.37	0.5520
Arisaema triphyllum	Ingestion	74	69	0.34	0.5617
Polygonum virginianum	Adhesion	21	17	0.22	0.6434
Mitchella repens	Ingestion	49	51	0.05	0.8247
Solidago flexicaulis	Wind	28	26	0.04	0.8456

thoughts; and to Dave Moeller, whose generous help facilitated the analyses. Their comments and those of Brian Barringer, Jesse Bellemare, Laurie Evanhoe, Monica Geber, Robert Harris, Shannon Murphy, Kim Sparks and Beatrijs Bossuyt also improved the manuscript. McIntire-Stennis and A.W. Mellon Foundation grants and a National Science Foundation graduate research fellowship provided funding.

References

Bellemare, J., Motzkin, G. and Foster, D.R. (2002) Legacies of the agricultural past in the forested present: an assessment of historical land-use effects on rich mesic forests. *Journal of Biogeography* 29, 1401–1420.

Bierzychudek, P. (1982) Life histories and demography of shade-tolerant temperate forest herbs: a review. *New Phytologist* 90, 757–776.

Bossuyt, B., Hermy, M. and Deckers, J. (1999) Migration of herbaceous plant species across ancient–recent forest ecotones in central Belgium. *Journal of Ecology* 87, 628–638.

Brunet, J. and von Oheimb, G. (1998) Migration of vascular plants to secondary woodlands in southern Sweden. *Journal of Ecology* 86, 429–438.

Colwell, R.K. (2000) EstimateS: statistical estimation of species richness and shared species from samples, version 6. http://viceroy.eeb.uconn.edu/estimates

Compton, J.E. and Boone, R.D. (2000) Long-term impacts of agriculture on soil carbon and nitrogen in New England forests. *Ecology* 81, 2314–2330.

Corbit, M.D. (1995) Hedgerows as habitat corridors for forest herbs in central New York, USA. MS thesis, Cornell University, Ithaca, New York.

Corbit, M., Marks, P.L. and Gardescu, S. (1999) Hedgerows as habitat corridors for forest herbs in central New York, USA. *Journal of Ecology* 87, 220–232.

Curtis, J.T. (1956) The modification of mid-latitude grasslands and forests by man. In: Thomas, W.C. (ed.) *Man's Role in Changing the Face of the Earth*. University of Chicago Press, Chicago.

Day, G.M. (1953) The Indian as an ecological factor in the northeastern forest. *Ecology* 34, 329–346.

Dupouey, J.L., Dambrine, E., Laffite, J.D. and Moares, C. (2002) Irreversible impact of past land use on forest soils and biodiversity. *Ecology* 83, 2978–2984.

Dzwonko, Z. and Loster, S. (1989) Distribution of vascular plant species in small woodlands on the Western Carpathian foothills. *Oikos* 56, 77–86.

Foster, D.R. (1992) Land-use history (1730–1990) and vegetation dynamics in central New England, USA. *Journal of Ecology* 80, 753–772.

Gotelli, N.J. and Colwell, R.K. (2001) Quantifying biodiversity: procedures and pitfalls in the measurement and comparison of species richness. *Ecology Letters* 4, 379–391.

Honnay, O., Bossuyt, B., Verheyen, K., Butaye, J., Jacquemyn, H. and Hermy, M. (2002) Ecological perspectives for the restoration of plant communities in European temperate forests. *Biodiversity and Conservation* 11, 213–242.

Marks, P.L. (2001) Reading the landscape #6: pastures of plenty. *Cornell Plantations Magazine* 56, 20–24.

Marks, P.L. and Gardescu, S. (1992) Vegetation of the central Finger Lakes region of New York in the 1790s. In: *Late Eighteenth Century Vegetation of Central and Western New York State on the Basis of Original Land Survey Records*. Bulletin No. 484, New York State Museum, Albany, New York, pp. 1–35.

Marks, P.L. and Gardescu, S. (2001) Inferring forest stand history from observational field evidence. In: Egan, D. and Howell, E.A. (eds) *The Historical Ecology Handbook: a Restorationist's Guide to Reference Ecosystems*. Island Press, Washington, DC, pp. 177–198.

Matlack, G.R. (1994) Plant species migration in a mixed-history forest landscape in eastern North America. *Ecology* 75, 1491–1502.

Matlack, G.R. (1997) Land use and forest habitat distribution in the hinterland of a large city. *Journal of Biogeography* 24, 297–307.

Neeley, J.A. (1965) *Soil Survey: Tompkins County, New York*. Series 1961, No. 25, USDA Soil Conservation Service, Government Printing Office, Washington, DC.

Nyland, R.D., Zipperer, W.C. and Hill, D.B. (1986) The development of forest islands in exurban central New York State. *Landscape and Urban Planning* 13, 111–123.

Peterken, G.F. and Game, M. (1984) Historical factors affecting the number and distribution of vascular plant species in the woodlands of central Lincolnshire. *Journal of Ecology* 72, 155–182.

Rackham, O. (1975) *Hayley Wood: its History and Ecology*. Cambridgeshire and Isle of Ely Naturalists' Trust, Cambridge, UK.

Rackham, O. (1980) *Ancient Woodland: its History, Vegetation and Uses In England*. Edward Arnold, London.

Singleton, R., Gardescu, S., Marks, P.L. and Geber, M.A. (2001) Forest herb colonization of postagricultural forests in central New York State, USA. *Journal of Ecology* 89, 325–338.

Smith, B.E., Marks, P.L. and Gardescu, S. (1993) Two hundred years of forest cover changes in Tompkins County, New York. *Bulletin of the Torrey Botanical Club* 120, 229–247.

Stover, M.E. and Marks, P.L. (1998) Successional vegetation on abandoned cultivated and pastured land in Tompkins County, New York. *Journal of the Torrey Botanical Society* 125, 150–164.

Vellend, M. (2003) Habitat loss inhibits recovery of plant diversity as forests regrow. *Ecology* 84, 1158–1164.

Vellend, M., Myers, J., Gardescu, S. and Marks, P.L. (2003) Dispersal of *Trillium* seeds by deer: implications for long-distance migration of forest herbs. *Ecology* 84, 1067–1072.

Verheyen, K., Honnay, O., Motzkin, G., Hermy, M. and Foster, D. (2003) Response of forest plant species to land use change in Europe and America: a life-history trait-based approach. *Journal of Ecology*, 91, 563–577.

Whitney, G.G. (1994) *From Coastal Wilderness to Fruited Plain: a History of Environmental Change in Temperate North America From 1500 to the Present*. Cambridge University Press, Cambridge.

Whitney, G.G. and Foster, D.R. (1988) Overstorey composition and age as determinants of the understorey flora of woods of central New England. *Journal of Ecology* 76, 867–876.

Whitney, G.G. and Somerlot, W.J. (1985) A case study of woodland continuity and change in the American midwest. *Biological Conservation* 31, 265–287.

Williams, M. (1989) *Americans and Their Forests: a Historical Geography*. Cambridge University Press, Cambridge, UK.

Wulf, M. (1997) Plant species as indicators of ancient woodland in northwestern Germany. *Journal of Vegetation Science* 8, 635–642.

Ancient Forests in Denmark and the Importance of *Tilia*

8

J.E. Lawesson

Institute of Biological Sciences, Aarhus University, Department of Systematic Botany, Risskov, Denmark

Tilia species are important constituents of natural broadleaved, deciduous forests of Europe, but have vanished from many forests during the past millennium. In this chapter, a survey of *Tilia* in Danish forests is presented and compared with forest management, age and structure. *Tilia cordata* and *T. platyphyllos* are mostly found in natural, ancient forests, and appear to be convenient indicators of forest management and forest age. Most populations are very small, but a few remarkable stands of both species with thousands of individuals are still surviving. The results point to the urgent need for installing active conservation and management measures, together with monitoring, in order to stabilize the most threatened populations of endangered woody species in Danish forests in general, and of *Tilia* species in particular.

8.1 Introduction

The world's forests cover about 40% of the ice-free land surface (52.4×10^6 km^2) at present, and account for a substantial portion of the earth's terrestrial carbon storage and biocomplexity (Waring and Running, 1995). Although a large fraction of forests have been converted into agricultural and urban areas, the remaining forests are of considerable importance for production of lumber, fuel wood and forestry-related products. Forests are important habitats for wildlife and for conservation of biodiversity (Honnay *et al.*, 1998, 1999a) and for providing recreational and aesthetic values.

Forests are major contributors to biospheric processes on all scales, from the modifying effects on sub-canopy microclimate to global regulation of CO_2 through source–sink dynamics (Kimmins, 1997). Forests strongly influence global climate, but are also much affected by global climate change through feedback mechanisms (Walter *et al.*, 2002). The reasons are therefore many why forests are among the favourite topics in science. In particular so-called natural and ancient forests have been much studied over the past decades, especially in Europe, where few and small fragments remain. Natural forests are valuable for the studies of natural processes and for use as reference points in monitoring of the environment. The identification of potential indicators of the naturalness of forests (Peterken, 1996) is therefore a high priority area of nature conservation (Noss, 1999).

   Tilia cordata and *T. platyphyllos* have been mentioned as important species in natural forests in Denmark (Böcher, 1946; Iversen, 1958; Møller, 1988; Lawesson, 2000) and Europe (Pigott, 1975, 1991; Falinski, 1986; Bobiec *et al.*, 2000) but few studies are known on the utility of *Tilia* species as indicators (Hermy *et al.*, 1999). We therefore would like to focus especially on *T. cordata* and *T. platyphyllos* in this chapter, and on their utility as indicators of natural, ancient forests in Denmark.

8.1.1 What does a natural broadleaved forest look like in northern Europe?

There has been much debate on the origins of the *Fagus* forests of northern Europe, due to their current dominance in large areas and their claimed natural conditions (Ellenberg, 1988; Peters and Poulson, 1994; Kuster, 1997). This discussion has been much biased from observations in strongly-managed forests of central Europe, poor in both structural and floristic heterogeneity and lacking small-scale dynamics. This misconception is now undergoing a rapid transition as new studies have identified the widespread coexistence of canopy tree species in natural species-rich deciduous forests of the temperate zone, and the important underlying role of disturbance regime and gap formation for the structural and floristic diversity in both ground and canopy layers (Bazzaz, 1996; Pancer-Koteja *et al.*, 1998; Sakio *et al.*, 2002). Species-rich mixed forests are well documented from several locations in northern and central Europe (Pigott, 1975; Falinski, 1986; Diekmann, 1994; Bobiec *et al.*, 2000; Lawesson, 2000) and appear to be the natural state of southern Scandinavian forests (Lawesson and Oksanen, 2002). This is also supported by the historical palaeoecological record, indicating that *Fagus sylvatica* became abundant in southern Scandinavian forests only during the past 3000 years (Björkman and Bradshaw, 1996, 1998; Bradshaw and Holmqvist, 1999). The changeover from mixed-canopy forests of Middle Holocene to those known today, dominated by *F. sylvatica*, has been attributed mainly to human large-scale disturbance (Iversen, 1958; Turner, 1962), such as increasing use of wood–pasture systems and forest clearance for agriculture, followed by abandonment of fields, which favours the replacement of *Tilia* with *Fagus* or *Quercus* (Lindbladh and Bradshaw, 1998; Bradshaw and Mitchell, 1999). Contrary to studies of pollen deposits of Flandrian age in Great Britain (Pigott, 1981), the details of prehistoric distribution of *T. cordata* and *T. platyphyllos* in Denmark are largely unknown, because *T. cordata* pollen have not been separated from *T. platyphyllos* in pollen diagrams, despite their discernible differences.

8.1.2 Indicators of ancient forest

Several herbaceous plant species, fungi, lichens and bryophytes are found only in ancient forests of Europe, and are used as indicators of a long forest continuity and original environmental conditions (e.g. Peterken, 1974; Tibell, 1992; Wulf, 1997; Lawesson *et al.*, 1998; Hermy *et al.*, 1999; Lindenmayer *et al.*, 2000; Sverdrup-Thygeson and Lindenmayer, 2003). Ancient forest indicators are usually absent from persistent seed banks, have low dispersal abilities and thus low colonization potential, and they are sensitive to large-scale anthropogenic disturbances (Rackham, 1980; Peterken and Game, 1981; Ellenberg, 1988) but may depend on small-scale natural disturbances such as gaps (Fischer *et al.*, 2002). Ancient forest indicators are absent or rare in new forests, as they may need centuries to colonize new habitats (Brunet *et al.*, 2000). Managed forests of northern Europe usually differ from ancient, natural forests in their age structure and physiognomy, spatial patterns of the tree layers, in low amounts of dead and decaying wood, and reduced number of natural gap-forming processes (Økland *et al.*, 2003). Indicator species are therefore important tools in nature conservation, as they combine both qualitative (forest quality) and quantitative (diversity) conservation criteria (Peterken, 1974, 1996), and thus may suggest the effects of change within a system, and how altered ecological processes affect populations. Indicator species have also received increasing attention for application in ecologically sustainable forest management (Lindenmayer, 1999). Woody species are less used as indicators of ancient forests (but see Rackham, 1980; Hermy *et al.*, 1999; Honnay *et al.*, 1999b; Mark and Lawesson, 2000).

8.1.3 Important tree species in natural broadleaved forests in northern Europe

When examining ancient forests in northern Europe, it may be worthwhile not only to use herbaceous plant species (or other groups of organism in the ground layers) as indicators, but also tree species, because they may possess some of the same traits mentioned above. In addition, tree species provide long-lasting evidence of former events, such as climatic changes, natural and man-made disturbances and successional processes (Schweingruber, 1996). Data obtained from tree populations are excellent for modelling of past, current and future trends (Loehle and LeBlanc, 1996). Understanding which tree species are convenient ancient forest indicators, and why, is therefore important for basic research as well as for the sound management of ancient forests, and for the restoration of exploited forests.

With the heterogeneous environment of natural, ancient forests, one should find early successional species (e.g. *Betula* spp., *Populus tremula*, *Prunus avium*, *Alnus glutinosa*), transitional species (*Fraxinus excelsior*, *Fagus sylvatica*, *Carpinus betulus*) as well as terminal species (e.g. *Quercus robur*, *T. cordata*, *T. platyphyllos*). Among these, the knowledge regarding *Tilia* is scant (but see Pigott, 1969, 1975, 1991), and the situation in Denmark is generally unknown (but see Wicksell, 1998).

8.1.4 Ancient forests in Denmark

In the Danish Forest Act, natural (ancient) forest is considered to be of long continuity, at least from the period around 1800, composed of local, naturally immigrated tree species, especially *T. cordata*. Natural forests in Denmark have probably all been influenced by humans, through logging, fuel wood collection and hunting. Planting or sowing of trees is not accepted in the Danish definition of natural forest. Ancient forests are here considered to be the remnant of natural forests that are known to have existed at least since the late 18th century, when the first extensive forest maps of Denmark were produced (Lawesson *et al.*, 1998). The ancient forests of today are often managed and influenced by forestry practice, but they remain, by definition, natural as long as their genetic contents are unchanged by humans. Much-impoverished natural forests are thus widespread, where selective logging and elimination of certain undesired tree species has been carried out. Such forests are still, according to Danish terms, considered to be natural, although they may be dominated by *F. sylvatica* or *Q. robur*. Smaller parts of the natural forests may be left without management, and may floristically and structurally resemble virgin forest, with natural regeneration, diverse age structure, many dead trees, and undisturbed soils. An increasing number of such so-called untouched forest patches have been liberated from normal forestry practice during the past decade.

8.1.5 Characteristics of *Tilia*

T. cordata is a nemoral to boreo-nemoral species with affinities to continental and eastern areas of Europe (Pochberger, 1967), and is widespread from

England and Norway eastwards to Siberia (Pigott, 1991). It is scattered in the lowlands further south, but common in the lower mountains of central Europe. *T. platyphyllos* is a demanding southern-nemoral thermophilic species with oceanic affinities (Rühl, 1968), with its northern range limit in Denmark. It is common in the mountain areas of central and southern Europe, while it only extends east to Ukraine. The two species often coexist in their sympatric range (Namvar and Spethmann, 1986, personal observation) and the putative hybrid, *T. europaea*, has been reported from such locations (Wicksell and Christensen, 1999). Both *Tilia* species are tall, terminal species in broadleaved nemoral forest, reaching a maximum height of 40 m, and *Tilia* may become very old, more than 1000 years (Pigott, 1981). *Tilia* will coppice readily and may survive for centuries in unfavourable conditions (Pigott, 1991), but will not produce flowers and pollen in dark and shady conditions. Only when the tree reaches the canopy or a gap, and is fully exposed to light, will flowers and fruit be produced (Pigott, 1991). Failing reproduction of *T. cordata* in Bialowiesa forest in Poland in the first half of 20th century has been ascribed exactly to this phenomenon by Pigott (1975). The seeds of *Tilia* mature slowly, and the germination rate from individuals in Denmark is low, especially for *T. cordata* (J.E. Lawesson, personal observation). Seeds are large and heavy, and although the diaspore is winged, the dispersal of seeds is poor, usually less than 100 m from the parent tree (Pigott, 1991; F. Borchsenius, personal communication). The seeds do not form persistent seed banks, but die after a few years (Barton, 1934). Both *Tilia* species thus need very specific conditions to establish, grow and reproduce (Pigott, 1989).

Several Scandinavian forest studies confirm that *Tilia* is a prominent component in several natural nemoral and boreo-nemoral forest types in Norway, Sweden and Denmark (Gram *et al.*, 1944; Böcher, 1946; Diekmann, 1994; Dierssen, 1996; Fremstad, 1997). Six major forest types with a predominance of *T. cordata* or *T. platyphyllos* were described by Lawesson (2000). These forests occur on both oligotrophic and eutrophic soils, and often contain *Q. robur*, *F. excelsior*, *F. sylvatica* and *Ulmus glabra* (Fig. 8.1, Fig. 8.2). *Tilia* forests are also found on rocky substrates on the isle of Bornholm in the Baltic Sea. *T. cordata* has a very wide ecological niche in Denmark, largely overlapping the niches of most other woody species (Lawesson and Oksanen, 2002).

8.2 Danish Forests and *Tilia*

8.2.1 Sampling forest characteristics and *Tilia* occurrences in Denmark

T. cordata and *T. platyphyllos* have both been little studied in Denmark (Ødum, 1968; Wicksell and Christensen, 1999; Lawesson, 2000). A more systematic survey has been undertaken in recent years as part of studies on relations between population ecology, climate change and genetic variability, and

Fig. 8.1. Ancient *Quercus robur–Tilia cordata* deciduous forest on mesotrophic sand. Holt Forest, Jutland, western Denmark, June 2001.

Fig. 8.2. Ancient *Fagus sylvatica–Tilia cordata* deciduous forest. Draved Forest, southern Jutland, Denmark, July 2001.

is presently in progress (see Acknowledgements). The main findings on distribution and stand characteristics of *Tilia* in Denmark are summarized here.

Tilia populations were surveyed in as many ancient forests as possible from 1998 onwards. We paid particular interest to the floristic and ecological context of *Tilia* communities (Lawesson, 2000), and forest and stand characteristics of the *Tilia* populations, such as age structure, reproductive success and niche characteristics. Based on the management type of the forests, they were divided into (i) *reserve*, and (ii) *commercial* forest. The overall structure of the forests was grouped into the following categories: (i) *high forest*; (ii) *low forest*; (iii) *coppice*; (iv) *coppice with high forest* (standards); and (v) *diverse forest*. The last category includes a forest with both tall and low forest, open areas and gaps, and may be present in managed forests, as well as in protected sites with some disturbance. The stand age of the area with *Tilia*, including all tree species, was estimated from diameter measurements, general knowledge of growth characteristics of the involved tree species and historical sources. The stands were divided into age categories, based on the oldest tree cohort present: *very old* (> 300 years), *old* (200–300 years), *mature* (100–200 years) and *young* (< 100 years). The size of *Tilia* populations was estimated through random walks and vegetation samples throughout the forests. The number of individuals (all size classes, including samplings) was divided into the following classes: *few* (< 50 individuals), *some* (50–100 individuals), *many* (101–500 individuals) and *abundant* (> 500 individuals). Independence between categories was tested with χ^2 in the programme S-PLUS (Anon., 1999).

8.2.2 *Tilia* populations in Denmark

Approximately 1100 forests and 5000 vegetation samples in Denmark have been surveyed, but *T. cordata* is only recorded in 77 forests and 169 sample plots. *T. cordata* populations are widespread in Denmark (Table 8.1, Fig. 8.3) but in the majority of cases (53 forests; 69%) the populations are small, with fewer than 50 individuals. In 21 forests the populations contain some or many individuals, e.g. on several small islands (e.g. Flatø, Lindø, Lille Vejlø), some small private forests (e.g. Kabbel Krat, Hamborg, Frederiksdal) and some state forests (e.g. Draved). Two reserves, i.e. Bolderslev in southern Jutland, and Ulvshale, in the isle of Møn, eastern Denmark, contain abundant quantities of *T. cordata*. Frejlev forest in south-eastern Denmark is the only commercial forest containing large quantities of *T. cordata*. Vindeholme Forest is another commercial forest, with virtually only *T. cordata* in the forest. However, this is the result of very long silvicultural management, where most other species have been eradicated. Therefore this forest was not included. There is a significant difference in population size categories between managed and commercial forests ($\chi^2 = 11.33$, df = 3, $P = 0.01$), but no relationship with stand age ($\chi^2 = 16.07$, df = 9, $P = 0.065$) or stand structure ($\chi^2 = 15.41$, df = 12, $P = 0.22$). There is a striking difference between vegetation plots with and without *T. cordata* (Table 8.2), with 172 and 293 species, respectively, and 127 species in common. The similarity between the two types of forest is very low

Table 8.1. Major Danish locations with natural *Tilia cordata* populations, ranked according to population size. For explanation see text.

Location	No.	Age	Structure	Management	Population size
Bolderslev	1	Mature	Diverse	Reserve	Abundant
Ulvshale	2	Old	High	Reserve	Abundant
Frejlev	3	Old	Coppice – high	Commercial	Abundant
Holt Krat	4	Young	Coppice – high	Commercial	Many
Flatø	5	Old	Coppice – high	Reserve	Many
Lindø, Maribo	6	Old	Diverse	Reserve	Many
Draved	7	Old	Diverse	Reserve	Many
Lindø	8	Very old	Diverse	Reserve	Many
Bobbeådal	9	Old	High	Commercial	Many
Kabbel Krat	10	Young	Low	Reserve	Many
Nørbjerg	11	Old	Low	Reserve	Many
Lille Vejlø	12	Young	High	Commercial	Many
Lundeborg	13	Old	High	Commercial	Many
Klåbygård	14	Old	High	Commercial	Many
Thurø Rev	15	Very old	High	Commercial	Many
Salne	16	Young	High	Commercial	Some
Hassel	17	Old	Coppice – high	Reserve	Some
Hamborg	18	Very old	High	Commercial	Some
Ekkodalen	19	Old	Low	Reserve	Some
Orenæs	20	Young	High	Commercial	Some
Frederiksdal	21	Mature	High	Commercial	Some
Grasten	22	Old	High	Commercial	Some
Jonstrup Vang	23	Old	High	Commercial	Some
Stensore	24	Old	High	Commercial	Some
Købelev	25	Mature	Coppice	Commercial	Few
Hønning Krat	26	Young	Coppice – high	Commercial	Few
Skårup Skovhaver	27	Young	Coppice – high	Commercial	Few
Toftlund	28	Young	Low	Commercial	Few
Skovbjerg Krat	29	Young	Low	Commercial	Few
Skovgårdslund	30	Young	Low	Commercial	Few
Krarup Lund	31	Old	Low	Commercial	Few
Borgø	32	Young	Diverse	Reserve	Few
æbelø	33	Very old	Diverse	Commercial	Few
Romsø	34	Young	High	Reserve	Few
Toft	35	Old	High	Commercial	Few
Bjæven	36	Old	High	Reserve	Few
Bognæs	37	Old	High	Commercial	Few
Helligdommen	38	Very old	High	Reserve	Few

Table 8.1. *Continued.*

Location	No.	Age	Structure	Management	Population size
Olstrup	39	Very old	High	Reserve	Few
Knudshoved	40	Young	Low	Reserve	Few
Ansager	41	Old	Low	Commercial	Few
Vang	42	Old	Low	Commercial	Few
Borre skov	43	Young	High	Commercial	Few
Beder	44	Young	High	Commercial	Few
Bøgebakke	45	Old	High	Commercial	Few
Glænø	46	Old	Coppice – high	Commercial	Few
Lystrup	47	Young	High	Commercial	Few
Roden	48	Young	High	Commercial	Few
Grenå	49	Young	High	Commercial	Few
Hjartbro	50	Young	High	Commercial	Few
Lillering	51	Young	High	Commercial	Few
Pamhule	52	Young	High	Commercial	Few
Rydhave	53	Young	High	Commercial	Few
Stensgaard	54	Young	High	Commercial	Few
Vindeby	55	Young	High	Commercial	Few
Jarrsskov	56	Young	High	Commercial	Few
Jelling skov	57	Young	High	Commercial	Few
Kollemorten Krat	58	Young	High	Commercial	Few
Nustrup	59	Young	High	Commercial	Few
Nørreskov	60	Young	High	Commercial	Few
Nørskovgaard	61	Young	High	Commercial	Few
Skovsgaard	62	Young	High	Commercial	Few
Skærsådalen	63	Young	High	Commercial	Few
Store Bøgeskov	64	Young	High	Commercial	Few
Store Fredskov	65	Young	High	Commercial	Few
Torpe	66	Young	High	Commercial	Few
Ravnholt	67	Young	High	Commercial	Few
Oren	68	Mature	High	Commercial	Few
Skovhaven	69	Mature	High	Commercial	Few
Ersted	70	Old	High	Commercial	Few
Fussingø	71	Old	High	Commercial	Few
Krabbesholm	72	Old	High	Commercial	Few
Krageskov	73	Old	High	Commercial	Few
Køge Ås	74	Old	High	Commercial	Few
Krenkerup	75	Very old	High	Reserve	Few
Vaar	76	Young	Low	Commercial	Few
Porskær Krat	77	Old	Low	Commercial	Few

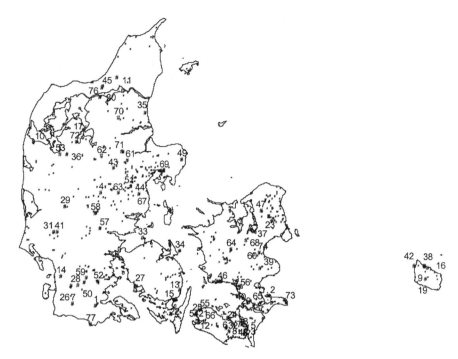

Fig. 8.3. Danish forest locations examined (dots) with indications of *Tilia cordata* populations (dots with numbers, according to Table 8.1). 1 cm = 50 km.

(Jaccard index: 0.23). Many ancient forest indicators (Hermy *et al.*, 1999) are present with higher frequencies in the forests with *T. cordata* than in those without (Table 8.2). This is the case both in the tree layer (e.g. *Acer campestre, Crateagus laevigata, Ilex aquifolium, Sorbus torminalis*) and herb layer (e.g. *Convallaria majalis, Luzula sylvatica, Ranunculus lanuginosus, Stellaria holostea*) while competitive, generalist species are more common in commercial forests (e.g. *Ranunculus ficaria, Mycelis muralis, Geranium robertianum, Galium odoratum* and many others).

T. platyphyllos* is less common than *T. cordata*, registered in only about 50 sample plots and 30 forests, mostly in southern Denmark (Table 8.3, Fig. 8.4) and most populations contain few individuals (22 forests, 73%). There is a significant difference in population size categories among the stand ages ($\chi^2 = 24.38$, df = 9, $P = 0.004$), with the largest populations restricted to one mature forest and four old or very old forest stands (five forests, 16.7% of all forests), while populations with few individuals are largely restricted to young forest stands (13 forests, 43.3%). The number of *T. platyphyllos* individuals does not depend on forest management ($\chi^2 = 5.13$, df = 3, $P = 0.16$), nor on stand structure ($\chi^2 = 12.77$, df = 12, $P = 0.38$) among the examined ancient forests.

Table 8.2.   Important tree and herb species in 168 *Tilia* forest plots and 168 plots without *Tilia* in the tree layer. Ancient forest species indicators according to Hermy *et al.* (1999) are marked with an asterisk.

	Tilia stands (frequency)	Other (frequency)
Trees		
*Acer campestre**	7	4
Carpinus betulus	9	8
*Corylus avellana**	15	8
*Crataegus laevigata**	13	4
Fagus sylvatica	39	40
Fraxinus excelsior	24	39
*Ilex aquifolium**	7	1
Populus tremula	13	4
Quercus robur	48	19
Sorbus aucuparia	26	8
*Sorbus torminalis**	3	
*Tilia cordata**	100	
Herbs		
*Anemone nemorosa**	45	50
*Carex sylvatica**	19	14
*Circaea lutetiana**	14	15
*Convallaria majalis**	28	4
*Dactylis glomerata**	18	14
*Lamiastrum galeobdolon**	11	8
*Luzula sylvatica**	9	0
*Melica uniflora**	33	32
*Mercurialis perennis**	11	8
*Milium effusum**	26	27
*Oxalis acetosella**	39	29
*Ranunculus lanuginosus**	2	0
*Stellaria holostea**	45	22
Deschampsia cespitosa	29	21
Deschampsia flexuosa	12	15
Galium odoratum	1	23
Geranium robertianum	6	11
Mycelis muralis	1	8
Ranunculus ficaria	1	8
Urtica dioica	8	16

8.3 Concluding Remarks

The study confirms that *T. cordata* and *T. platyphyllos* are mostly found in natural, ancient forests, and appear to be convenient indicators of forest management and forest age, respectively, which supports other studies where habitat

Table 8.3. Major Danish locations with natural *Tilia platyphyllos* populations, ranked according to population size. For explanations, see text.

Location	No.	Age	Structure	Management	Population size
Bolderslev	1	Mature	Diverse	Reserve	Abundant
Flatø	2	Old	Coppice – high	Reserve	Many
Frejlev	3	Old	Coppice – high	Commercial	Many
Lundeborg	5	Old	High	Commercial	Many
Suserup	4	Very old	Diverse	Reserve	Many
Hamborg	6	Very old	Diverse	Commercial	Some
Nørholm	8	Young	High	Commercial	Some
Stensgaard	7	Young	High	Commercial	Some
Corselitze	27	Old	High	Commercial	Few
Grenå	18	Young	High	Commercial	Few
Helnæs	9	Young	Coppice	Commercial	Few
Hostrupskov	16	Young	High	Commercial	Few
Jerslev	17	Young	High	Commercial	Few
Kabbel krat	30	Young	Low	Reserve	Few
Krageskov	15	Old	High	Commercial	Few
Krenkerup	12	Very old	Diverse	Reserve	Few
Merritskov	19	Young	High	Commercial	Few
Nykøbing	22	Young	High	Commercial	Few
Næset	14	Very old	Diverse	Reserve	Few
Nørskovgaard	23	Young	High	Commercial	Few
Pramskov	24	Young	High	Commercial	Few
Ryslinge kohave	25	Young	High	Commercial	Few
Rådmandshave	13	Very old	Diverse	Reserve	Few
Sinebjerg	20	Young	High	Commercial	Few
Stensore	21	Old	High	Commercial	Few
Strærup	10	Young	Coppice – high	Commercial	Few
Thurø	29	Very old	High	Commercial	Few
Toftlund	28	Mature	High	Commercial	Few
Tolne	26	Very old	High	Commercial	Few
Æbelø	11	Young	Diverse	Reserve	Few

diversity and forest age are key factors in species-rich forests (Dumortier *et al.*, 2002). Due to their causal relationships with forest management and continuity (Pigott, 1969; Peterken and Jones, 1987), both *Tilia* species seem to be good indicators of ecological continuity (Angelstam and Pettersson, 1997) over long time scales, i.e. several centuries, which is probably the natural generation time for the dominant and subdominant species (*Tilia* spp., *Q. robur*, *F. sylvatica*) in natural, ancient forests of Denmark (Lawesson and Oksanen, 2002). Such ancient forest indicators may need several hundreds of years of ecological continuity due to dependence on certain microhabitat structures

Fig. 8.4. Danish forest locations examined (dots) and indications of *Tilia platyphyllos* locations (dots with numbers, according to Table 8.2). 1 cm = 50 km.

that need many years to develop (e.g. canopy gap formation, exposure of mineral soil, addition of logs as a regeneration substrate). Explanations for such dependence on ecological continuity include poor dispersal and slow rates of establishment (Sverdrup-Thygeson and Lindenmayer, 2003), as, indeed, found with both *Tilia* species. *Tilia* is able to survive vegetatively, as shown by its widespread but low-abundant occurrence in many Danish forests, where small populations have been known for centuries (Ødum, 1968). However, *Tilia* spp. can only reproduce by seeds and thus consolidate their role, when growing in forests where environmental conditions provide sufficient light gaps for the formation of flower buds and seeds, and probably also soil moisture for the seeds to germinate and grow (Eisenhut, 1959). Such conditions are rare in commercial forests and young, shading tree stands. The two *Tilia* species appear separately, and together in their sympatric range, to be very good indicators of long forest continuity. The simultaneous presence of several other ancient forest indicators with *Tilia* populations (Table 8.2) supports this.

The incidence of *T. cordata* (Table 8.1) in Bolderslev, Ulvshale and Frejlev forests appears to rank these forests as outstanding, with a high degree of naturalness. Bolderslev forest in south-western Denmark is a 150 ha unmanaged forest, with large populations of both *Tilia* species and a heterogeneous forest canopy. Ulvshale forest in eastern Denmark is a well-known ancient forest with large populations of rare species, such as *T. cordata*, *S. torminalis*, *C. betulus*, *Acer platanoides* and others (Böcher, 1946). Frejlev forest, in south-eastern Denmark, is a large managed forest, but with many different owners. Draved forest is also included in the list, as containing many *T. cordata* individuals, some of them very old. This is in accordance with former reports on the unique character of this forest (e.g. Iversen, 1958).

T. platyphyllos indicates a high degree of naturalness in two of the above forests (Bolderslev, Frejlev), but also in, for example, Lundeborg, Hamborg and Suserup, all located in eastern Denmark, and only the latter protected as a private reserve. Suserup forest has very old trees, especially of *F. sylvatica* and *Q. robur*. However, the tree flora is rather poor, and the population of *T. platyphyllos* is mainly located in one sector of the forest, and has been claimed to have been planted some centuries ago (Friis Møller, personal communication). Despite reports of the ancient, natural character of Suserup forest, mainly based on its physiognomic characteristics (Emborg, 1998), it appears less genuine than several other ancient forests of Denmark, mentioned above.

In order to reveal the functional and causal linkages between *Tilia* populations and species richness, habitat quality and other animal and plant species, further studies are needed. A particularly important component of the testing and validation of potential indicator species is the examination of patterns of coexistence between indicator species and other species or processes for which they are supposed to be indicative (Lindenmayer, 1999). This should involve rigorous statistical testing of whether suspected indicator species do, in fact, indicate certain environmental or biological conditions, and this would include modelling of niche properties (Lawesson and Oksanen, 2002). As the importance of a given factor varies with the scale of the study, the value of the indicator species must be evaluated at the spatial and temporal scales, where this is relevant (Auerbach and Shmida, 1987; Dumortier *et al.*, 2002).

The fact that *Tilia* populations are found in so few Danish forests while, without any reasonable doubt, they had a much wider occurrence earlier, stresses the urgent need for installing active conservation and management measures, together with monitoring, in order to fulfil the Convention on Biological Diversity (United Nations, 1992) and stabilize the most-threatened populations of endangered woody species in Danish forests in general, and of *Tilia* species in particular. Notably, the selection of forest reserves or protected areas in Denmark is not based on empirical knowledge of the population size, age structure and functional role of woody indicators such as *Tilia* spp., *A. platanoides*, *A. campestre*, *Taxus baccata*, *I. aquifolium* or *S. torminalis*. Instead, *ad hoc* selection (Pressey, 1994) of forest areas, largely with the same herbaceous species composition, has left many woody species without protection. Paradoxically, despite the fact that long-term monitoring of natural areas, including forests, has been stressed by many (Lindenmayer, 1999; Lawesson *et al.*, 2000), effective forest-monitoring programmes focusing on the ecology and dynamics of the main component of the forests, the trees, have yet to be installed in Denmark. Thus, most of the populations of *T. cordata* and *T. platyphyllos* in ancient Danish forests have no legal protection, and could vanish at any time, if private or public owners decide to recreate the assumed open forests or parkland of the past (cf. Vera, 2000), or want to convert the ancient forests to monocultures of *Abies, Fagus* or *Quercus*. It is therefore recommended, as done in boreal forest (Sverdrup-Thygeson and Lindenmayer, 2003) that natural *Tilia* populations are used as one of the prime criteria for selection of forest sites for conservation, such as reserves, national parks and areas regulated under the EU Habitats Directive (Lund, 2002).

Acknowledgements

Most of the work presented here was done while I was a guest researcher and supervisor at the Institute of Biological Sciences, Aarhus University, in collaboration with the groups of associate professor Finn Borchsenius and Professor Henrik Balslev, Aarhus University; Associate Professor Mads C. Forchhammer, Zoological Institute, University of Copenhagen and Professor Carol Baskin, Lexington, Kentucky. The work benefited considerably from the dissertations of several MSc students, for whom I was the technical supervisor. Invaluable information and hints on *Tilia* populations in Denmark were furthermore put at my disposal by Ulla Wicksell, Rune Kristiansen, Ole Klitgaard, Henrik Staun, Peter F. Møller, Jens Cristian Svenning, Kåre Fog and many others. I am deeply thankful to them all. This study was supported by the project 'Scaling the interface between climate and ecology: long-term dynamics in plants and animals across natural and cultivated landscapes in Denmark' 2000–2002, supported by the Danish Natural Science Research Council (DRC) to M.C. Forchhammer. The study also received financial support from the Carlsberg Foundation and DRC towards the project: 'Ecology, life history traits and genetic variation in two *Tilia* species and their hybrid in Denmark – the response to a changing environment?'

References

Angelstam, P. and Pettersson, B. (1997) Principles of present Swedish forest biodiversity management. *Ecological Bulletin* 46, 191–203.

Anonymous (2001) *S-Plus 6 Guide to Statistics 1–2*, Insightful Corporation, Seattle, Washington, DC.

Auerbach, M. and Shmida, A. (1987) Spatial scale and the determinants of plant species richness. *Trends in Ecology and Evolution* 2, 238–242.

Barton, V. (1934) Dormancy in *Tilia* seeds. *Contribution of the Boyce Thomson Institute* 6, 69–89.

Bazzaz, F.A. (1996) *Plants in Changing Environments. Linking Physiological, Population, and Community Ecology*. Cambridge University Press, Cambridge, UK.

Björkman, L. and Bradshaw, R.H.W. (1996) The immigration of *Fagus sylvatica* and *Picea abies* into a natural forest stand in southern Sweden during the last 2000 years. *Journal of Biogeography* 23, 235–244.

Björse, G. and Bradshaw, R.H.W. (1998) 2000 years of forest dynamics in southern Sweden: suggestions for forest management. *Forest Ecology and Management* 104, 15–26.

Bobiec, A., van der Burgt, H., Meijer, K., Zuyderduyn, C., Haga, J. and Vlaanderen, B. (2000) Rich deciduous forests in Bialowieza as a dynamic mosaic of developmental phases: premises for nature conservation and restoration management. *Forest Ecology and Management* 130, 159–175.

Böcher, T.W. (1946) Vegetationsstudier på Halvøen Ulvshale. *Botanisk Tidsskrift* 46, 1–42.

Bradshaw, R.H.W. and Holmqvist, B.H. (1999) Danish forest development during the last 3000 years reconstructed from regional pollen data. *Ecography* 22, 53–62.

Bradshaw, R.H.W. and Michell, F.J.G. (1999) The palaeoecological approach to reconstructing former grazing–vegetation interactions. *Forest Ecology and Management* 120, 3–12.

Brunet, J., von Oheimb, G. and Diekmann, M. (2000) Factors influencing vegetation gradients across ancient-recent woodland borderlines in southern Sweden. *Journal of Vegetation Science* 11, 515–524.

Diekmann, M. (1994) Deciduous forest vegetation in Boreo-nemoral Scandinavia. *Acta Phytogeographica Suecica* 80, 1–112.

Dierssen, K. (1996) *Vegetation Nordeuropas*. Eugen Ulmer, Stuttgart, Germany.

Dumortier, M., Butaye, J., Jacquemyun, H., van Camp, N., Lust, N. and Hermy, M. (2002) Predicting vascular plant species richness of fragmented forests in agricultural landscapes in Central Belgium. *Forest Ecology and Management* 158, 85–102.

Eisenhut, G. (1959) Keimung der Lindenfrüchte. *Forstwissenschaftliches Centralblatt* 14, 47–56.

Ellenberg, H. (1988) *Vegetation Ecology of Central Europe*, 4th edn. Cambridge University Press, Cambridge, UK.

Emborg, J. (1998) Understorey light conditions and regeneration with respect to the structural dynamcis of a near-natural temperate deciduous forest in Denmark. *Forest Ecology and Management* 106, 83–95.

Falinski, J.B. (1986) *Vegetation Dynamics in Temperate Lowland Primeval Forests. Ecological Studies in Bialowieza Forest*. Dr W. Junk Publishers, Dordrecht, The Netherlands.

Fischer, A., Lindner, M., Abs, C. and Lasch, P. (2002) Vegetation dynamics in central European Forest Ecosystems (near-natural as well as managed) after storm events. *Folia Geobotanica* 37, 17–32.

Fremstad, E. (1997) *Vegetationstyper i Norge*. NINA Temahefte 12, Trondheim, Norway.

Gram, K., Jørgensen, C.A. and Køie, M. (1944) De jyske Egekrat og deres Flora. *Det Kongelige Danske Videnskabernes Selskab, Biologiske Skrifter* 3, 1–210.

Helliwell, D.R. (1989) Lime trees in Britain. *Arboricultural Journal* 13, 119–123.

Hermy, M., Honnay, O., Firbank, L., Grashof-Bokdam, C.J. and Lawesson, J. (1999) An ecological comparison between Ancient and other forest plant species of Europe, and the implications for forest conservation. *Biological Conservation* 91, 9–22.

Honnay, O., Degroote, B. and Hermy, M. (1998) Ancient-forest plant species in western Belgium: a species list and possible ecological mechanisms. *Belgian Journal of Botany* 130, 139–154.

Honnay, O., Hermy, M. and Coppin, P. (1999a) Impact of habitat quality on forest plant species colonization. *Forest Ecology and Management* 115, 157–170.

Honnay, O., Hermy, M. and Coppin, P. (1999b) Effects of area, age and diversity of forest patches in Belgium on plant species richness, and implications for conservation and reforestation. *Biological Conservation* 87, 73–84.

Iversen, J. (1958) Pollenanalytischer Nachweis des Reliktencharacters eines Jutischen Linden-Mischwaldes. *Veroff. Des Geobotanischen Institute Rubel in Zurich* 33, 137–144.

Kimmins, J.P. (1997) *Forest Ecology. A Foundation for Sustainable Management*, 2nd edn. Prentice Hall, Upper Saddle River, New Jersey.

Kuster, H. (1997) The role of farming in the postglacial expansion of beech and hornbeam in the oak woodlands of central Europe. *The Holocene* 7, 239–242.

Lawesson, J.E. (2000) Danish deciduous forest types. *Plant Ecology* 151, 199–221.

Lawesson, J.E. and Oksanen, J. (2002) Niche characteristics of Danish woody species as derived from coenoclines. *Journal of Vegetation Science* 13, 279–290.

Lawesson, J.E., De Blust, G., Grashof, C., Firbank, L., Honnay, O., Hermy, M., Hobitz, P. and Jensen, L.M. (1998) Species diversity and area-relationships in Danish beech forests. *Forest Ecology and Management* 106, 235–245.

Lawesson, J.E., Eilertsen, O., Diekmann, M., Reinikainen, A., Gunnlaugsdottir, E., Fosaa, A.M., Carøe, I., Skov, F., Groom, G., Økland, T., Økland, R., Andersen, P.N. and Bakkestuen, V. (2000) A concept for vegetation studies and monitoring in the Nordic countries. *TemaNord* 2000, 517.

Lindbladh, M. and Bradshaw, R.H.W. (1998) The origin of present forest composition pattern in southern Sweden. *Journal of Biogeography* 25, 463–477.

Lindenmayer, D.B. (1999) Future directions for biodiversity conservation in managed forests: indicator species, impact studies and monitoring programs. *Forest Ecology and Management* 115, 277–287.

Lindenmayer, D.B., Margules, C.R. and Botkin, D.B. (2000) Indicators of biodiversity for ecologically sustainable forest management. *Conservation Biology* 14, 941–950.

Loehle, C. and LeBlanc, D. (1996) Model-based assessments of climate change effects on forests: a critical review. *Ecological Modelling* 90, 1–31.

Lund, M.P. (2002) Performance of the species listed in the European Community 'Habitats' Directive as indicators of species richness in Denmark. *Environmental Science and Policy* 5, 105–112.

Mark, S. and Lawesson, J.E. (2000) Plants as indicators of the conservation value of Danish beech forests. *Proceedings IAVS Symposium* 1998, 158–161.

Møller, P.F. (1988) *Naturskov i Statsskovene*. DGU Intern rapport 4. Ministry of the Environment, Copenhagen.

Namvar, K. and Spethmann, W. (1986) The indigenous forest tree species of the genus *Tilia*. *Allgemeine Forstzeitschrift* 3, 42–45.

Noss, R.F. (1999) Assessing and monitoring forest biodiversity: a suggested framework and indicators. *Forest Ecology and Management* 115, 135–146.

Ødum, S. (1968) Udbredelsen af træer og buske in Danmark. *Botanisk Tidsskrift* 64, 5–118.

Økland, T., Rydgren, K., Økland, R.H., Storaunet, K.O. and Rolstad, J. (2003) Variation in environmental conditions, understorey species number, abundance and composition among natural and managed *Picea abies* forest stands. *Forest Ecology and Management* 177, 17–37.

Pancer-Koteja, E., Szagrzyk, J. and Bodziarczyk, J. (1998) Small-scale spatial pattern and size structure of *Rubus hirtus* in a canopy gap. *Journal of Vegetation Science* 9, 755–762.

Peterken, G.F. (1974) A method of assessing woodland flora for conservation using indicator species. *Biological Conservation* 6, 239–245.

Peterken, G.F. (1996) *Natural Woodland. Ecology and Conservation in Northern Temperate Regions.* Cambridge University Press, Cambridge, UK.

Peterken, G.F. and Game, M. (1981) Historical factors affecting the distribution of *Mercurialis perennis* in Central Lincolnshire. *Journal of Ecology* 69, 781–796.

Peterken, G.F. and Jones, E.W. (1987) Forty years of change in Ladypark Wood; the old growth stands. *Journal of Ecology* 75, 477–512.

Peters, R. and Poulson, T.L. (1994) Stem growth and canopy dynamics in a world-wide range of *Fagus* forests. *Journal of Vegetation Science* 5, 421–432.

Pigott, C.D. (1969) The status of *Tilia cordata* and *T. platyphyllos* on the Derbyshire limestone. *Journal of Ecology* 57, 491–504.

Pigott, C.D. (1975) Natural regeneration of *Tilia cordata* in relation to forest-structure in the forest of Bialowiesa, Poland. *Philosophical Transactions of the Royal Society of London*, Series B 75, 151–179.

Pigott, C.D. (1981) The status, ecology and conservation of *Tilia platyphyllos* in Britain. In: Synge, H. (ed.) *The Biological Aspects of Rare Plant Conservation.* John Wiley & Sons, Chichester, UK, pp. 305–317.

Pigott, C.D. (1989) Factors controlling the distribution of *Tilia cordata* at the northern limits of its geographical range. 4. Estimated ages of the trees. *New Phytologist* 112, 117–121.

Pigott, C.D. (1991) Biological flora of the British Isles: *Tilia cordata* Miller. *Journal of Ecology* 79, 1147–1207.

Pochberger, J. (1967) *Die Verbreitung der Linde Insbesondere in Oberösterreich.* Mitteilungen der Forstlichen Bundes-Versuchsanstalt, Vienna.

Pressey, R.L. (1994) *Ad hoc* reservations: forward or backward steps in developing representative reserve systems. *Conservation Biology* 8, 662–668.

Rackham, O. (1980) *Ancient Woodland, its History, Vegetation and Uses in England.* Arnold, London.

Rühl, A. (1968) Lindenmischwälder im südlichen Nordwestdeutschland. *Allgemeine Forst – und Jagdzeitung* 139, 118–130.

Sakio, H., Kubo, M., Shimano, K. and Ohno, K. (2002) Coexistence of three canopy tree species in a riparian forest in the Chichibu mountains, Central Japan. *Folia Geobotanica* 37, 45–61.

Schweingruber, F.H. (1996) *Tree Rings and Environment. Dendroecology.* Swiss Federal Institute for Forest, Snow and Landscape Research, Bern, Switzerland.

Sverdrup-Thygeson, A. and Lindenmayer, D.B. (2003) Ecological continuity and assumed indicator fungi in boreal forest: the importance of the landscape matrix. *Forest Ecology and Management* 174, 353–363.

Tibell, L. (1992) Crustose lichens as indicators of forest continuity in boreal coniferous forest. *Nordic Journal of Botany* 12, 427–450.

Turner, J. (1962) The *Tilia* decline: an anthropogenic interpretation. *New Phytologist* 70, 328–341.

United Nations (1992) *Convention on Biological Diversity*. United Nations, New York.

Vera, F.W.M. (2000) *Grazing Ecology and Forest History*. CAB International, Wallingford, UK.

Walter, G.-R., Post, E., Convey, P., Menzel, A., Parmesan, C.A., Beebee, T.J.C., Fromentin, J.-M., Hoegh-Guldberg, O. and Bairlein, F. (2002) Ecological responses to recent climate change. *Nature* 416, 389–395.

Waring, R.H. and Running, S.W. (1995) *Forest Ecosystems*, 2nd edn. Academic Press, San Diego, California.

Wicksell, U. (1998) *En Foreløbig Opgørelse over Lindeforekomster I Danmark*. Arbejdsrapportserien 18–2002. Forskningscentret for Skov og Landskab, Hørsholm, Denmark.

Wicksell, U. and Christensen, K.I. (1999) Hybridization among *Tilia cordata* and *T. platyphyllos* (Tiliaceae) in Denmark. *Nordic Journal of Botany* 19(6), 673–684.

Wulf, M. (1997) Plant species as indicators of ancient woodland in northwestern Germany. *Journal of Vegetation Science* 8, 635–642.

Ecology and Population Genetics of *Carabus problematicus* in Flanders, Belgium: is Forest History Important?

9

K. Desender,[1] E. Gaublomme,[1] P. Verdyck,[1,2] H. Dhuyvetter[1] and B. De Vos[3]

[1]Department Entomology, RBINSc, Brussels, Belgium; [2]University of Antwerp, Groenenborgerlaan, Antwerp, Belgium; [3]Institute for Forestry and Game Management, AMINAL, Geraardsbergen, Belgium

The large, wingless ground beetle, *Carabus problematicus*, occurs in western Europe from Finland to southern France, including the British Isles. Belgium is situated near the centre of its geographical distribution area, but this does not imply that the species occurs equally distributed over the country. Allozyme and microsatellite markers allowed a preliminary population genetic comparison between 11 populations of *C. problematicus*, collected in different eco-regions. Genetic diversity is higher both in larger and ancient forests, as compared to a recent forest. In general, the results show a high structuring for both allozyme and microsatellite data. Nearly all populations can be characterized genetically. An important conclusion is that conservation genetic values can differ dramatically for the same species according to the eco-region, and this on a relatively small geographic scale. *C. problematicus* thus shows a remarkable interaction between eco-region and forest history, habitat preference and population genetics. At least in Flanders, it appears to be a powerful model species, with a wide range of applications for conservation studies and genetic monitoring at different levels of perception.

9.1 Introduction

Today, most woodland in the region of Flanders (i.e. the northern part of Belgium) is highly fragmented or degraded, mainly due to excessive human interference in the past. Flemish forests are among the most fragmented in western Europe, with a long and well-documented fragmentation history, characterized by periods of regression and expansion (Bloemers and Van Dorp, 1991; Tack *et al.*, 1993; Tack and Hermy, 1998). A first regression started with the introduction of neolithic cultures in the area. The processes of forest degradation and disappearance continued and reached a first peak during the Roman period, as a result of large-scale agricultural exploitation. During the early medieval period, woodland recovered to some extent, while from the high Middle Ages onwards renewed deforestation took place (Verhulst, 1990). From then onwards, gradual deforestation was only occasionally interrupted by renewed cultivation of woodland. During the 19th century, a final large-scale deforestation took place in Flanders. As a result, forests now cover less than 10% of the total surface, and only a relatively small part can be characterized as ancient forest. In Flanders, ancient forest is defined as being older than 225 years, and continuously present on a given location since the maps of Count de Ferraris, about 1775 (see also Tack *et al.*, 1993; Tack and Hermy, 1998).

As a consequence, many forest organisms are expected to have disappeared or to have been highly influenced during historical times in Flanders (e.g. Tack *et al.*, 1993; Ervynck *et al.*, 1994; Bossuyt *et al.*, 1999; Desender *et al.*, 1999; Hermy *et al.*, 1999; Honnay *et al.*, 1999a,b). In addition to forest area reduction, loss of typical habitat and decrease of habitat quality, many organisms are supposed to have suffered increasingly from a long history of isolation between forest fragments. This seems especially true for taxa or species with a low dispersal power (for plants, cf. Hermy *et al.*, 1999; for invertebrates, cf. Desender *et al.*, 1999). Such organisms are no longer able to disperse easily from old fragments, where they survived until today, to other, more recently restored forest sites. We can therefore hypothesize that such species are now strong indicators for the historical ecology of forests in Flanders.

Ground beetles (Coleoptera, Carabidae) are a speciose family of invertebrates (with *c.* 400 species in Belgium), showing a number of characteristics of high interest for studies on woodland history. They mostly show a high species richness and several species have a pronounced (micro-) habitat preference for forest interiors. The recent and historical occurrence of ground beetles is very well documented in Flanders – derived from a large amount of distribution data since about 1850 (Desender *et al.*, 1994), as well as from archaeological data (Ervynck *et al.*, 1994; Desender *et al.*, 1999). A documented Red Data Book was published some years ago for this region (Desender *et al.*, 1995). Moreover, in these beetles, dispersal power and gene flow are reflected in the development of flight wings and of indirect flight muscles. Species are either full-winged, wingless or wing polymorphic (Den Boer *et al.*, 1980; Desender, 1989), and can even show variability in flight muscle development (Desender, 2000). Most of the typical woodland species are constantly wingless or never develop functional flight musculature (Desender, 1989; Assmann, 1999; Desender *et al.*, 1999).

In recent years, carabid diversity, distribution and population ecology, as well as population genetics, have been investigated in Flanders, within the context of habitat fragmentation and historical ecology. Recently, ecological assemblages of ground beetles derived from a large number of forests have been documented (Desender *et al.*, 2002b). They strongly suggest that forest history, as well as recent ecological characteristics, are both important in structuring these assemblages. At the population level, a variety of carabid model species has now been studied in detail, with emphasis on beetles from forests and saltmarshes (Desender and Turin, 1989; Desender *et al.*, 1998, 1999). Recent investigations on woodland ground beetles aim at evaluating the role of historical and present-day ecology and forest fragmentation (area and isolation) by studying ecological and genetic differentiation and diversity. Such studies may throw light on the mechanisms (other than habitat quality decline or loss of suitable habitat) responsible for the loss of forest species during woodland fragmentation (e.g. genetic erosion and reduced gene flow). The results may increase our understanding of the actual conservation value of woodlands and suggest future priorities for woodland restoration. Until now, population genetic studies from Flanders forest beetles have used mainly allozyme data, including a variety of ground beetle species with known distribution, ecology and dispersal power (e.g. Desender *et al.*, 1999, 2002a). *Abax ater*, a rather eurytopic forest species, showed relatively modest genetic differentiation (Desender *et al.*, 1999). *Carabus auronitens*, another model species, showed extreme genetic differentiation and local adaptation in relation to forest historical ecology (Desender *et al.*, 2002a). As this carabid has survived until now in only very few forests in Flanders, there was a growing need to study additional, but relatively more common, forest species.

In this chapter we present ecological as well as preliminary genetic data on another forest ground beetle, *C. problematicus*. This large and constantly wingless beetle occurs in western Europe from Finland to southern France, including the British Isles (Turin, 2000). In Belgium, situated near the centre of its current geographic distribution area, there appears to be remarkable diversity in abundance, as well as in preferred (micro-) habitat of this beetle, according to the eco-region (Desender, unpublished). Here, we will evaluate whether forest history (ancient versus recent forests) and forest area are both important for the ecology, distribution and genetics of *C. problematicus* populations in Flanders. This will be interpreted against the background of different eco-regions. Only preliminary genetic data are given here.

9.2 Material and Methods

Detailed sampling of the carabid yearly cycle has been performed in many habitats in Flanders, mainly since 1985 (Desender *et al.*, 1995), including recent efforts in a large number of forests (Desender *et al.*, 2002b). Some 100 forests, distributed over the entire region of Flanders and differing in area and history, have been documented in detail. Several replicate sites were sampled. In eastern and western Flanders, 12 recent and 27 ancient forests were sampled. Twenty forests from Brabant, including three recent ones, were studied, while in the Campine region, 33 forests, including seven ancient woodlands, were investigated. Sampling was performed by means of continuous pitfall trapping (three traps (glass jam jars of diameter 9.5 cm) per site, 5 m apart, on a line; emptied at fortnightly intervals) for at least 1 complete year at a given location. Field sampling for the population genetic studies mainly took place during 2000. In this chapter, only data on the presence or absence of *C. problematicus* in each of the studied forests are used. The data were compared to forest area and forest history, calculated or derived from maps in an ArcView GIS (Environmental Systems Research Institute, 1992). Detailed electronic forest data were available from the 'Digital version of the forest reference layer' (MVG, LIN, AMINAL, Dept Bos en Groen, edition 2001, OC-GIS Flanders).

Genetic variability and differentiation were studied by means of allozymes and microsatellite DNA markers. Three hundred and fifty beetles were collected from 11 populations (from ten forests, including one population from the extreme southern part of Belgium as an external reference). Figure 9.1 shows the positions of the sampling locations with reference to the geographical distribution of *C. problematicus* in the country. One large ancient forest in Flanders (Sonian) was sampled at two different sites.

Variation at enzyme loci was studied by cellulose acetate electrophoresis (Hebert and Beaton, 1989). After a pilot study in which 27 enzyme loci were screened, six polymorphic enzymes (aspartate aminotransferase (AAT), glucose-6-phosphate (G6PDH), isocitrate dehydrogenase (IDH1, IDH2), peptidase-Z (PEPZ) and glucose-6-phosphate isomerase (GPI)) were retained for further analysis. We used two legs of each beetle (in 50 μl of distilled water), while the rest of the body was kept in ethanol for future DNA extraction or for morphometric studies. The same populations were also screened for four variable microsatellite markers, developed recently (Gaublomme *et al.*, 2003) in cooperation with INRA (Centre de Biologie et de Gestion des Populations, Montferrier-sur-Lez, France). [For more details on field sampling, electrophoresis and software for standard population genetic analyses, see Desender *et al.* (1998) and Desender and Verdyck (2001).] In addition, analysis of molecular variance (AMOVA) was performed using ARLEQUIN 2 (Schneider *et al.*, 2000), in order to partition and test the genetic variance between eco-regions, between populations within regions and within populations. The population genetic data based on the polymorphic loci showed no deviations from

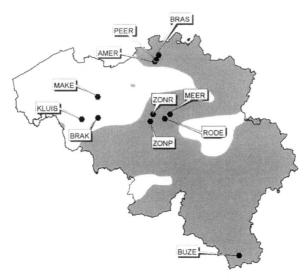

Fig. 9.1. Localization of forests where samples for the population genetic study were gathered (black circles). The background map shows the geographical distribution of *Carabus problematicus* in Belgium (shaded area).

Hardy–Weinberg equilibrium and no linkage disequilibrium, which means that the studied enzyme, as well as microsatellite loci, could be used as independent markers.

9.3 Results and Discussion

9.3.1 Geographical distribution and habitat preference of *C. problematicus* in different eco-regions

Figure 9.2 shows the entire geographical area of *C. problematicus* and its detailed distribution in Belgium. Figure 9.3 compares the number of ancient and recent forests occupied by the species in different eco-regions of Flanders and for different forest size classes. In the south (Fig. 9.2, ellipse A), especially in the forested regions of the Ardennes, *C. problematicus* occurs abundantly in many woodlands, with a preference for acid soils (Baguette, 1993). In the north, in the Campine region, the species is also common (Fig. 9.2, ellipse B), but here it is found increasingly (in low numbers) in open heath-like habitats as well as in forests, mainly on acid sandy soil (Desender, unpublished data). This corresponds to the general trend of its habitat preference shifting to more open habitat types towards northern Europe (Rijnsdorp, 1980; Houston, 1981) or in cooler mountain areas at high altitude (Sparks *et al.*, 1995; Butterfield, 1996; Turin, 2000; Buse *et al.*, 2001; Dennis *et al.*, 2002). A comparison of our recent

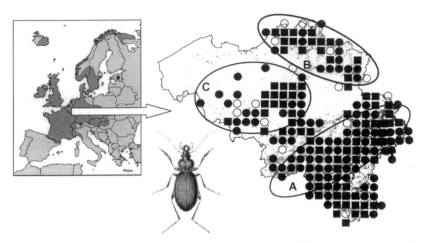

Fig. 9.2. Distribution range of *Carabus problematicus*: full range (map inset, after Turin *et al.*, 2003) and compilation of data from Belgium. Universal Transeverse Mercator (UTM) 10 km squares: open symbols, data before 1950; black circles, data since 1950; black squares, data from both time periods. See text for a further explanation of the added ellipses A–C.

pitfall data shows that in the Campine region, 24 out of 26 recent forests, and all but the smallest of seven ancient forests, are occupied by the species, in all forest area classes. Apparently, the beetle colonized many recent forests in that area, rather successfully, probably through its sporadic occurrence in heath-like habitats in between forests.

However, in the central and westernmost parts of Flanders, *C. problematicus* is increasingly rare (Fig. 9.2, ellipse C) and is a strong indicator of ancient forests here. *C. problematicus* is absent, without exception, from all 15 recent forests studied so far, while only 11 out of 44 ancient forests (nearly all larger than 100 ha) are occupied by the species (Fig. 9.3). Archaeological data show that *C. problematicus* once occurred at other sites in that region (Desender *et al.*, 1999). The species probably disappeared mainly as a result of the negative consequences of forest fragmentation.

To conclude, there is a very pronounced and remarkable interaction between forest history and the actual occurrence of *C. problematicus* according to the eco-region in Flanders. As such, this beetle is a powerful model species for ecological as well as genetic studies, offering a wide array of possibilities for studies at different levels of perception; for example, between ancient forests of different size, between different eco-regions, and, for the Campine region, between forests of different ages. From a conservation perspective, the importance of forest history for the occurrence of this wingless beetle thus differs dramatically according to the eco-region. These topics are further explored below, in terms of population genetics.

9.3.2 Genetic diversity and differentiation of *C. problematicus* in relation to forest history and area in Flanders

Figure 9.4 compares genetic diversity estimates (mean number of alleles and gene diversity) between allozyme and microsatellite data, in ancient forests and in one recent forest. Table 9.1 shows the results of the correlation analyses between genetic diversity estimates of *C. problematicus* and forest area and population size (forest area multiplied by beetle density, i.e. the relative estimate based on sampling over 1 year).

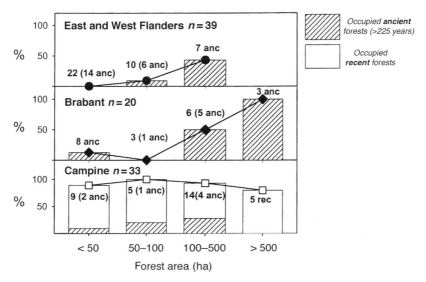

Fig. 9.3. Forest occupancy by *Carabus problematicus* in recent and ancient forests of different sizes in three regions of Flanders.

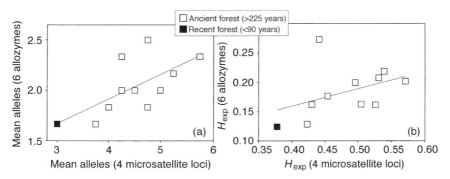

Fig. 9.4. Genetic diversity compared between allozyme and microsatellite loci in ten ancient, and a single recent, population of *Carabus problematicus* in Flanders. (a) Mean number of alleles per locus; (b) gene diversity estimates (= H_{exp} or expected heterozygosity).

Table 9.1. Pearson correlation between genetic diversity estimates of
C. problematicus and forest area or forest area × beetle density.

	Mean number of alleles		Gene diversity (H_{exp})	
	Allozymes	Microsatellites	Allozymes	Microsatellites
Forest area	0.45	0.46	0.26	**0.68**
Area × beetle density	**0.69**	0.47	0.42	0.52

Relative estimate based on sampling for 1 year; significant values in bold, $P < 0.05$.

Overall, we obtained a similar pattern of population genetic diversity
estimates for allozymes and microsatellites, based on the mean number of
alleles (Fig. 9.4a) or gene diversity (i.e. expected heterozygosity, Fig. 9.4b).
Without exception, the single recent forest population studied showed the
lowest genetic diversity; in other words, a higher genetic diversity is observed
in older forests (older beetle populations). Highest values were obtained for the
largest ancient forests (mainly from the Brabant region, on loamy soil). Forest
area as well as population size (relative measure) were both positively, in several cases significantly, correlated with genetic diversity estimates (Table 9.1).
Larger populations (i.e. forests) thus indicate a higher genetic diversity.

These preliminary results clearly suggest that genetic diversity differs,
according to the eco-region, in conjunction with differences in forest age, area
and isolation. We are now in the process of adding more population genetic
data from other forests and regions, in order to test the hypotheses suggested
here on a firmer basis. Although *C. problematicus* occurs in many recent forests
in the northern Campine region (see above), its genetic diversity seems to be
lower there, suggesting that colonization occurred through a few individuals
only (founder events). Whether this could cause long-term problems for the
evolutionary potential of the species in that region (Frankham *et al.*, 2002)
remains an open question.

The relatively small dataset on *C. problematicus* already shows high structuring: genetic differentiation between the 11 populations studied amounts to
an overall estimate of the proportion of genetic variation among populations
(*Fst*; six allozymes) of 0.076 ($P < 0.0001$) and an overall *Rst* (an alternative
estimate regularly used for microsatellite data; four microsatellites) of 0.106
($P < 0.0001$). This means that as much as about 10% of the total genetic
variation can be attributed to differences between populations, a high value
in comparison to other organisms studied earlier in this respect, certainly in
comparison to the relatively small geographical scale of our study area (Ward
et al., 1992). Two-by-two tests of genetic differentiation show that nearly all
populations can be characterized genetically. A general AMOVA (Table 9.2)
between the populations of the three eco-regions in Flanders shows highly
significant genetic variation between regions, between populations within
regions and within populations.

Table 9.2. AMOVA results based on all enzyme and microsatellite loci of 10 populations of *C. problematicus* in three eco-regions in Flanders (Campine, Brabant and East and West Flanders).

Source of variation	df	Sum of squares	Variance components	Percentage of variation	Significance
Among eco-regions	2	45.138	0.08331	4.68	$P < 0.004$
Among populations within regions	7	41.550	0.07124	4.00	$P < 0.000$
Within populations	596	969.458	1.62661	91.32	$P < 0.000$
Total	605	1056.147	1.78115		

Genetic results for the rather eurytopic forest beetle species *A. ater* (Desender *et al.*, 1999) have shown significant differentiation (based on allozymes) between populations (as a suggested result of reduced gene flow) on a relatively large spatial scale, whereas genetic erosion could not (yet) be observed for this species. Another model beetle species, *C. auronitens*, nowadays extremely rare in Flanders, has been studied both at enzyme loci and by means of microsatellite markers (Desender *et al.*, 2002a). Results in that species showed extreme genetic differentiation and isolation by mechanisms other than geographical distance (isolation by non-distance) in relation to forest historical ecology. The comparison of results from both types of genetic markers suggested that genetic drift (loss of genetic variability in small populations), reduced gene flow (as a consequence of increased isolation), as well as local adaptation in conjunction with forest history, could explain the observed patterns of genetic variation. The preliminary genetic results on *C. problematicus*, given in this chapter, also show a relatively high amount of genetic differentiation but here, genetic structure as well as diversity appears to depend on, or interact with, eco-region and forest age, as well as forest size. On the one hand, the strong influence of forest history could be a consequence of the extreme degree of forest fragmentation that took place in the major parts of Flanders. On the other hand, it could have resulted from the very long history of this fragmentation. Certainly these aspects have important regional conservation genetic implications.

As a more general conclusion, the study of different model species can suggest different patterns and processes which have acted upon the ecology, genetic structure and diversity of organisms as a possible outcome of forest fragmentation and forest history. In north-western and central Europe *C. problematicus* is rather abundant in many regions, and almost nowhere endangered (Assmann, 1999; Turin *et al.*, 2003). However, at least in many parts of Flanders, this is not the case. Another important conclusion from our study on *C. problematicus* is therefore that conservation (genetic) values can differ dramatically for the same species according to the region, and this on a relatively small geographic scale. In Flanders, *C. problematicus* therefore appears to be a powerful model species, with a wide range of applications for conservation studies and genetic monitoring at different levels of perception.

Acknowledgements

Regional and national nature conservation and forestry authorities allowed us to sample invertebrates in their forests. J.-P. Maelfait [Instituut Natuurbehoud (IN)], D. De Bakker and F. Hendrickx [Universiteit Gent (RUG)], S. Thys and L. De Bruyn [Universitair Centrum Antwerpen (RUCA), IN], K. Vandekerkhove and D. Vandenmeersschaut [Instituut Bosbouw en Wildbeheer (IBW)], D. Maddelein and D. Gorissen [Administratie Milieu Natuur en Landschap (AMINAL)], A. Drumont, V. Choquet, J. Constant and V. Versteirt [Royal Belgian Institute of Natural Sciences (RBINSc)], and J.Y. Rasplus (INRA, Montferrier-sur-Lez, France) are acknowledged for their help and support in various ways. Many graduate students contributed with enthusiasm to the sorting of tens of thousands of forest invertebrates. H. Turin kindly delivered a pre-print map of the distribution of *C. problematicus*.

This study was supported partly by the RBINSc (Department of Entomology) and through financial support by the Flemish Government (AMINAL, Dept. Bos en Groen, projects B&G/15/96 and B&G/29/98; AMINAL/Natuur/ project VLINA 00/015) and the Federal Government [Federale Diensten voor Wetenschappelijke, Technische en Culturele Aangelegenheden (DWTC), project MO 36/006]. The investigations were also carried out within the framework of the Flemish Research Network FWO.010.97N ('Ecological genetics: patterns and processes of genetic variation in natural populations'). P. Verdyck is a postdoctoral researcher of the Fund for Scientific Research – Flanders.

References

Assmann, T. (1999) The ground beetle fauna of ancient and recent woodlands in the lowlands of north-west Germany (Coleoptera, Carabidae). *Biodiversity and Conservation* 8, 1499–1517.

Baguette, M. (1993) Habitat selection of carabid beetles in deciduous woodlands of southern Belgium. *Pedobiologia* 37, 365–378.

Bloemers, J.H.F. and Van Dorp, A. (1991) *Pre- and Protohistorie van de Lage Landen.* Open Universiteit, De Haan, The Netherlands.

Bossuyt, B., Hermy, M. and Deckers, J. (1999) Migration of herbaceous plant species across ancient–recent forest ecotones in central Belgium. *Journal of Ecology* 87, 628–638.

Buse, A., Hadley, D. and Sparks, T. (2001) Arthropod distribution on an alpine elevational gradient: the relationship with preferred temperature and cold tolerance. *European Journal of Entomology* 98, 301–309.

Butterfield, J. (1996) Carabid life-cycle strategies and climate change: a study on an altitude transect. *Ecological Entomology* 21, 9–16.

Den Boer, P.J., Van Huizen, T.H.P., Den Boer-Daanje, W., Aukema, B. and Den Bieman, F.M. (1980) Wing polymorphism and dimorphism in ground beetles as stages in an evolutionary process (Coleoptera: Carabidae). *Entomologia Generalis* 6, 107–134.

Dennis, P., Aspinall, R.J. and Gordon, I.J. (2002) Spatial distribution of upland beetles in relation to landform, vegetation and grazing management. *Basic and Applied Ecology* 3, 183–193.

Desender, K. (1989) *Dispersievermogen en Ecologie van Loopkevers (Coleoptera, Carabidae) in België: een Evolutionaire Benadering.* Koninklijk Belgisch Instituut voor Natuurwetenschappen, Brussels.

Desender, K. (2000) Flight muscle development and dispersal in the life cycle of carabid beetles: patterns and processes. *Bulletin van het Koninklijk Belgisch Instituut voor Natuurwetenschappen, Entomologie* 70, 13–31.

Desender, K. and Turin, H. (1989) Loss of habitats and changes in the composition of the ground- and tiger beetle fauna in four West-European countries since 1950 (Coleoptera: Carabidae, Cicindelidae). *Biological Conservation* 48, 277–294.

Desender, K. and Verdyck, P. (2001) Geographic scaling and genetic differentiation in two highly mobile European saltmarsh beetles. *Belgian Journal of Zoology* 131, 29–40.

Desender, K., Dufrêne, M. and Maelfait, J.-P. (1994) Long term dynamics of carabid beetles in Belgium: a preliminary analysis on the influence of changing climate and land use by means of a database covering more than a century. In: Desender, K., Dufrêne, M., Loreau, M., Luff, M.L. and Maelfait, J.-P. (eds) *Carabid Beetles: Ecology and Evolution.* Kluwer Academic Publishers, Dordrecht, The Netherlands, pp. 247–252.

Desender, K., Maes, D., Maelfait, J.-P. and Van Kerckvoorde, M. (1995) Een gedocumenteerde Rode lijst van de zandloopkevers en loopkevers van Vlaanderen. *Mededelingen van het Instituut voor Natuurbehoud* 1, 1–208.

Desender, K., Backeljau, Th., Delahaye, K. and De Meester, L. (1998) Age and size of European saltmarshes and the population genetic consequences for ground beetles. *Oecologia* 114, 503–513.

Desender, K., Ervynck, A. and Tack, G. (1999) Beetle diversity and historical ecology of woodlands in Flanders. *Belgian Journal of Zoology* 129, 139–156.

Desender, K., Verdyck, P., Gaublomme, E., Dhuyvetter, H. and Rasplus, J.-Y. (2002a) Extreme genetic differentiation and isolation by non-distance in *Carabus auronitens* in relation to historical ecology of forests in Flanders (Belgium). In: Szyszko, J. *et al.* (eds) *How to Protect or What We Know About Carabid Beetles.* Warsaw Agricultural University Press, Warsaw, pp. 227–235.

Desender, K., De Bakker, D., Versteirt, V. and De Vos, B. (2002b) A baseline study on forest ground beetle diversity and assemblages in Flanders (Belgium). In: Szyszko, J. *et al.* (eds) *How to Protect or What We Know About Carabid Beetles.* Warsaw Agricultural University Press, Warsaw, pp. 237–245.

Environmental Systems Research Institute (1992) *ArcView Geographical Information System,* Version 3.2. Environmental Systems Research Institute, Redlands, California.

Ervynck, A., Desender, K., Pieters, M. and Bungeneers, J. (1994) Carabid beetles as palaeo-ecological indicators in archaeology. In: Desender, K., Dufrêne, M.,

Loreau, M., Luff, M.L. and Maelfait, J.-P. (eds) *Carabid Beetles: Ecology and Evolution*. Kluwer Academic Publishers, Dordrecht, The Netherlands, pp. 261–266.

Frankham, R., Ballou, J.D. and Briscoe, D.A. (2002) *Introduction to Conservation Genetics*. Cambridge University Press, Cambridge, UK.

Gaublomme, E., Dhuyvetter, H., Verdyck, P., Mondor-Genson, G., Rasplus, J.-Y. and Desender, K. (2003) Isolation and characterisation of microsatellite loci in the ground beetle *Carabus problematicus* (Coleoptera, Carabidae). *Molecular Ecology Notes* 3(3), 341–343.

Hebert, P.D.N. and Beaton, M.J. (1989) *Methodologies for Allozyme Analysis Using Cellulose Acetate Electrophoresis*. Helena Laboratories, Beaumont, Texas.

Hermy, M., Honnay, O., Firbank, L., Grashof-Bokdam, C. and Lawesson, J.E. (1999) An ecological comparison between ancient and other forest plant species of Europe, and the implications for forest conservation. *Biological Conservation* 91, 9–22.

Honnay, O., Hermy, M. and Coppin, P. (1999a) Effects of area, age and diversity of forest patches in Belgium on plant species richness, and implications for conservation and reforestation. *Biological Conservation* 87, 73–84.

Honnay, O., Hermy, M. and Coppin, P. (1999b) Impact of habitat quality on forest plant species colonization. *Forest Ecology and Management* 115, 157–170.

Houston, W.W.K. (1981) The life cycles and age of *Carabus glabratus* Paykull and *C. problematicus* Herbst (Col.: Carabidae) on moorland in northern England. *Ecological Entomology* 6, 263–271.

Rijnsdorp, A.D. (1980) Pattern of movement in and dispersal from a Dutch forest of *Carabus problematicus* Hbst. (Coleoptera, Carabidae). *Oecologia (Berl.)* 45, 274–281.

Schneider, S., Roessli, D. and Excoffier, L. (2000) *ARLEQUIN Version 2.00: a Software Programme for Population Genetics Data Analysis*. Genetics and Biometry Laboratory, University of Geneva, Italy.

Sparks, T.H., Buse, A. and Gadsden, R.J. (1995) Life strategies of *Carabus problematicus* (Coleoptera, Carabidae) at different altitudes on Snowdon, North Wales. *Journal of Zoology, London* 236, 1–10.

Tack, G. and Hermy, M. (1998) Historical ecology of woodlands in Flanders. In: Kirkby, K.J. and Watkins, C. (eds) *The Ecological History of European Forests*. CAB International, Wallingford, UK, pp. 283–292.

Tack, G., Van den Bremt, P. and Hermy, M. (1993) *Bossen van Vlaanderen. Een Historische Ecologie*. Davidsfonds, Leuven, Belgium.

Turin, H. (2000) *De Nederlandse Loopkevers, Verspreiding en Oecologie (Coleopter: Carabidae)*. Nederlandse Fauna 3. Nationaal Natuurhistorisch Museum Naturalis, KNNV Uitgeverij and EIS Nederland, Leiden, The Netherlands.

Turin, H., Penev, L. and Casale, A. (2003). *The Genus* Carabus *L. in Europe. A Synthesis*. Pensoft Publishing, Sofia, Bulgaria.

Verhulst, A. (1990) *Précis d'Histoire Rurale de la Belgique*. Vrije Universiteit Brussel, Brussels.

Ward, R.D., Skibinski, D.O.F. and Woodwark, M. (1992) Protein heterozygosity, protein structure, and taxonomic differentiation. In: Hecht, M.K., Wallace, B. and Macintyre, R.J. (eds) *Evolutionary Biology*. Plenum Press, New York, pp. 73–159.

Colonization of Oak Plantations by Forest Plants: Effects of Regional Abundance and Habitat Fragmentation

10

J. Brunet

Southern Swedish Forest Research Centre, Swedish University of Agricultural Sciences, Alnarp, Sweden

This study aimed at analysing how colonization of forest plants is affected by source population size and seed dispersal type in recent oak plantations with different degrees of isolation. Colonization was studied in 40 oak plantations on former arable fields in the Torup-Skabersjö area, southern Sweden. Twenty-two oak stands in ancient woodland were used as reference areas. Species richness increased linearly with stand age in plantations adjacent to ancient woodland. The oldest adjacent plantations (70–80 years) approached a species richness similar to that of the ancient woodlands. The results indicate that interspecific differences in colonization rates within contiguous forest landscapes are largely determined by the size of local source populations. Species richness of plantations decreased with increasing distance from ancient woodlands. There was no increase of species richness between stands of age 40–80 years in isolated plantations. Species that regularly colonized isolated plantations often had adhesive seeds or were ferns with wind-dispersed spores. In landscapes with highly fragmented forest patches, interspecific differences in the ability of long-distance dispersal seem to control colonization patterns to a large extent.

10.1 Introduction

Ecologists have long focused on local biotic and abiotic ecological factors to explain the distribution patterns of plant species. The possibility that succession could be controlled by interspecific differences in dispersal capacity remained largely unexplored.

More recently, dispersal limitation has been recognized as an important ecological factor controlling forest species distributions. Most deciduous forest plants lack a persistent soil seed bank and have to migrate to restored sites (Thompson *et al.*, 1998). In a pioneer study, Peterken and Game (1984) found that many forest plants had a low ability to colonize isolated recent woodlots in a highly fragmented landscape. Later work at landscape scale confirmed that pattern in mixed history forest landscapes in Europe and eastern North America (Wittig *et al.*, 1985; Whitney and Foster, 1988; Dzwonko and Loster, 1992; Matlack, 1994; Petersen, 1994; Wulf, 1997; Honnay *et al.*, 1998; Graae and Sunde, 2000; Bellemare *et al.*, 2002).

However, studies of secondary woods directly adjacent to ancient woodland have shown that most forest plants are able to colonize secondary stands, even though unfavourable soil or micro-climate may inhibit recruitment in some cases (Dzwonko, 1993, 2001; Matlack, 1994; Brunet and von Oheimb, 1998a,b; Bossuyt *et al.*, 1999; Honnay *et al.*, 1999; Singleton *et al.*, 2001; Verheyen and Hermy, 2001a,b).

Several recent studies demonstrate that distance to suitable source populations is a generally important factor controlling colonization of new stands. However, logistic regression analyses showed highly variable, species-specific changes in probability of occurrence in relation to distance from source populations (Grashof-Bokdam and Geertsema, 1998; Bossuyt *et al.*, 1999; Butaye *et al.*, 2001; Verheyen and Hermy, 2001a,b).

Interspecific differences may arise during all stages of the colonization process: (i) seed source limitation due to low production of viable seeds; (ii) dissemination (dispersal) limitation due to low amount of seeds dispersed, short-distance dispersal or patchy seed dispersal; and (iii) establishment limitation due to unfavourable site conditions, competition or seed predation (Jordano and Godoy, 2002).

The aim of this chapter is to compare the relative importance of seed source limitation (estimated as source population size) and dispersal limitation (estimated as seed dispersal type) for colonization of isolated and non-isolated recent woodlands by forest vascular plants. The basic assumption is that colonization of non-isolated recent woodlands is mainly controlled by seed source limitation, whereas differences in seed dispersal largely control colonization of isolated woodlands.

10.2 Material and Methods

10.2.1 Study sites

The study was conducted in oak (*Quercus robur*) stands within the 1000 ha forest landscape at the estates of Skabersjö and Torup in the province of Skåne, southernmost Sweden (55°32′N, 13°11′E). The study area is situated in the nemoral vegetation zone. Deciduous hardwood forests (class *Querco-Fagetea*) on dystric and eutric cambisols typically dominate on well-drained sites. The mean annual temperature is *c.* 7.5°C, mean annual precipitation *c.* 600 mm. Between 1920 and 1960, and again from 1990 onwards, the estate foresters of Skabersjö have established a large number of oak stands on old fields. These plantations are unique for Sweden, concerning the number of stands and ranges of both stand age (7–84 years) and spatial isolation from colonization sources (0–2000 m), and offer an excellent opportunity to analyse long-term colonization processes by forest plants under relatively homogeneous site conditions.

Using historical maps and forest management plans, the history of each oak stand was determined. For the Torup estate, detailed cadastral survey maps were available from the years 1694, 1799 and 1915, and forest management plans for 1945, 1973, 1983, 1992 and 2001. For the Skabersjö estate, cadastral maps were available from 1733 (part of the area), 1769–1771 (part of the area) and 1915, and forest management plans from 1839, 1878, 1926–1929, 1942, 1981 and 1999. Furthermore, the year of plantation is known for most stands.

Areas that had been wooded continuously since the oldest survey maps were considered to be *ancient woodland*. Here, both former wood pastures and wood meadows are included. In the study area, most ancient woodlands were characterized by beech (*Fagus sylvatica*) forests that have persisted for the past 300 years. Especially on eutric cambisols, oak–hazel (*Q. robur, Corylus avellana*) stands have persisted in some areas. Forest stands that had developed on former open pastures, meadows or cultivated fields were classified as *recent woodland*. In this chapter, plant colonization was only studied in recent woodlands on former cultivated fields, where I assume that typical woodland species have not survived, either as plants or seeds.

According to the historical data, stands of the following categories were chosen for further study (cf. Table 10.1):

1. As reference areas, containing the regional pool of oak forest species, ten stands of semi-natural ancient woodland, all containing veteran oak trees, and 12 oak stands that have been planted within ancient woodland were used. Mean species richness was not significantly different between these stand types (24.1 versus 24.9 species, $P = 0.5$). They were therefore merged in further analyses and are referred to as ancient woodlands.

2. Plant colonization was studied in 40 stands on old fields, of which 37 had a canopy dominated by *Q. robur*, two by *Betula pendula* and one by *Fraxinus excelsior* (I assume that no forest species have survived *in situ* during the period of cultivation):

- 16 plantations were situated in a large contiguous forest in the eastern part of the study area close to ancient woodlands (further referred to as *adjacent plantations*);

Table 10.1.   Stand age, stand area and species richness of typical forest plants
(means and ranges) in different types of oak woodland in the Torup–Skabersjö area,
Skåne, south Sweden.

Stand type	Number of stands	Means (ranges) of stand age (years)	Means (ranges) of stand area (ha)	Means (ranges) of species richness (number)
Ancient woodland	22	–	2.31a (0.5–10.5)	24.5a (19–31)
Adjacent plantations	16	63.9a (38–82)	1.36a (0.5–3.4)	18.6b (14–24)
Isolated plantations	12	58.8a (43–84)	1.29a (0.6–3.1)	8.8c (5–12)
Young plantations	12	9.6b (7–12)	3.95b (1.3–9.6)	1.9d (0–8)

Different letters indicate significant differences according to ANOVA and
Tukey-test ($P < 0.05$).

- 12 stands are found in small secondary forest fragments in the western
  part of the study area, where woodlands are embedded in a matrix of
  cultivated fields (further referred to as *isolated plantations*). Stand age
  between these two groups was not significantly different (Table 10.1); and
- the remaining 12 oak plantations were established between 1990 and
  1995 and are treated as a separate group referred to as *young plantations*.
  All but one of the young plantations are adjacent to older woodland, but
  only two of them have direct contact with ancient woodland. The young
  plantations were significantly younger and larger than the stands of the
  other categories (Table 10.1).

10.2.2 Field survey

A floristic survey was conducted during 2002 in all 62 stands listed above. For
comparison, all other deciduous stands in ancient woodland in the study area
($n = 118$, total area 345 ha) were also surveyed. Vernal species were studied
between 3 April and 22 May 2002, summer species were surveyed between
27 June and 1 August 2002. Each deciduous forest stand was surveyed for the
occurrence of typical forest floor plants, i.e. shade-tolerant herbs, graminoids
and ferns (cf. Weimarck and Weimarck, 1985; see also nomenclature).
Transect walks through each stand were taken until *c.* 10 min had passed
without detection of new species.

   The abundance of each forest species in a stand was estimated using a
three-graded ordinal scale: 1, cover < 1% of stand area; 2, cover 1–20% of
stand area; 3, cover > 20% of the stand area. Further, the stand borders were
examined to map species restricted to the stand margin. These were assigned a
value of 0.5.

10.2.3 Data analysis

Differences in species frequency between different stand types were tested with the 2-way contingency chi-squared test. Relations between stand age, distance to nearest ancient woodland and species richness were analysed with linear and stepwise regression. Species richness of different stand types was compared using ANOVA and the Tukey-test (Zar, 1996).

For each species, the following estimates of regional population sizes were calculated, based on data from the 22 ancient woodland stands (source populations) and from recent plantations (colonizing populations):

- species frequency (% occurrence in stands of different categories); and
- mean value of cover % estimates (class 0.5 was assigned 0.5% cover; class 1, 1% cover; class 2, 5% cover; class 3, 25% cover).

Both frequency and mean cover of woodland species were closely correlated between ancient oak stands ($n = 22$) and all deciduous stands in ancient woodland in the study area ($n = 140$, $r^2 = 0.931$ for frequency, $r^2 = 0.929$ for mean cover, both $P < 0.0001$). Thus, frequency and cover of forest species in ancient oak stands was a good indicator of overall size of source populations. The relationship between source population sizes and colonizing population sizes was analysed with linear regression.

10.3 Results

10.3.1 Species richness

In total, 48 forest species were found during the survey, of which 47 species occurred in ancient stands, 36 in adjacent plantations, 27 species in isolated plantations and 12 species in young plantations (Table 10.2). Species frequency showed a pronounced bimodal distribution in the ancient stands. Fourteen species showed frequencies > 90% and 12 species were found in < 10% of the ancient oak stands. A similar pattern, though less pronounced, occurred in adjacent plantations. Mean species richness was higher in ancient woodlands than in all types of plantations (Table 10.1). However, the oldest adjacent plantations approached a species richness similar to that of the ancient woodlands (Fig. 10.1). Species richness increased linearly with stand age in adjacent plantations ($n = 16$, $r^2 = 0.710$, $P = 0.0001$; regression including young and adjacent plantations: $n = 28$, $r^2 = 0.932$, $P < 0.0001$). No such trend was observed in isolated plantations ($n = 12$, $r^2 = 0.028$, $P < 0.5$; Fig. 10.1).

Species richness of plantations decreased with increasing distance from ancient woodlands ($n = 28$, $r^2 = 0.531$, $P = 0.001$, excluding young plantations). Species richness was not significantly related to stand area in isolated

Table 10.2.   Frequency (%) of forest plants in ancient oak stands ($n = 22$), older oak plantations adjacent to ancient woodland ($n = 16$), older isolated oak plantations ($n = 12$) and young oak plantations ($n = 12$).

	Ancient woodlands	Adjacent plantations	Isolated plantations	Young plantations	Diaspore dispersal
Colonize all plantations					
Milium effusum	100	88	83	8*	**adhesive**
Circaea lutetiana	91	94	67	8*	**adhesive**
Poa nemoralis	91	94	100	25*	**adhesive**
Ranunculus ficaria	91	88	75	8*	ants
Dryopteris filix-mas	86	94	75	50*	wind
Festuca gigantea	86	94	100	8*	**adhesive**
Adoxa moschatellina	64	81	42	0*	ingested
Dryopteris carthusiana	59	63	50	33	wind
Colonize adjacent plantations					
Anemone nemorosa	100	94	33*	0*	ants
Stellaria holostea	100	100	17*	17*	none
Stellaria nemorum	100	88	42*	8*	ingested
Oxalis acetosella	95	100	8*	0*	**auto**
Viola reichenbachiana	95	100	25*	8*	ants
Athyrium filix-femina	91	88	33*	8*	wind
Stachys sylvatica	73	44	8*	0*	adhesive
Galium odoratum	59	38	17*	0*	**adhesive**
Elymus caninus	55	38	0*	0*	adhesive
Convallaria majalis	50	25	8*	0*	ingested
Gagea spathacea	50	31	0*	0*	ants
Corydalis intermedia	36	31	0*	0*	ants
Pulmonaria obscura	36	25	0*	0*	ants
Generally slow colonizers					
Melica uniflora	95	69*	8*	0*	**ants**
Mercurialis perennis	95	63*	33*	8*	ants
Gagea lutea	91	25*	0*	0*	ants
Lamium galeobdolon	91	56*	8*	0*	ants
Scrophularia nodosa	86	50*	8*	0*	**none**
Maianthemum bifolium	77	19*	8*	0*	ingested
Polygonatum multiflorum	73	6*	8*	0*	ingested
Anemone ranunculoides	59	0*	0*	0*	ants
Rare species					
Campanula trachelium	23	6	0	0	wind
Trientalis europaea	23	0*	0	0	none
Epipactis helleborine	18	25	8	0	wind
Brachypodium sylvaticum	14	19	8	0	**adhesive**
Carex sylvatica	14	0	0	0	**ants**

Table 10.2. *Continued.*

	Ancient woodlands	Adjacent plantations	Isolated plantations	Young plantations	Diaspore dispersal
Gymnocarpium dryopteris	14	0	0	0	wind
Hedera helix	9	6	0	0	ingested
Hepatica nobilis	9	0	0	0	ants
Lathraea squamaria	9	0	0	0	ants
Melampyrum pratense	9	0	0	0	ants
Paris quadrifolia	9	6	8	0	ingested
Allium ursinum	5	0	0	0	ants
Bromus benekenii	5	6	0	0	adhesive
Campanula latifolia	5	0	0	0	none
Chrysosplenium alternifol.	5	0	0	0	**none**
Melica nutans	5	0	0	0	ants
Pyrola minor	5	0	0	0	none
Sanicula europaea	5	6	0	0	**adhesive**
Actaea spicata	0	6	0	0	ingested

Significantly lower frequencies in plantations as compared to ancient stands are indicated with * (chi-squared statistic, $P < 0.05$). Dispersal type according to the literature (cf. Hermy *et al.*, 1999). Dispersal type in bold face indicates evidence of adhesive seed dispersal by dog, roe deer or wild boar (see discussion for references).

plantations ($n = 12$, $r^2 = 0.044$, $P = 0.5$), but there was a trend of increasing species richness with stand area in adjacent plantations ($n = 16$, $r^2 = 0.182$, $P = 0.0964$). In a stepwise regression analysis of all 40 plantations, with stand age, stand area and distance from the nearest ancient woodland as independent variables, stand age was selected first ($r^2 = 0.645$, $P < 0.0001$), followed by distance to ancient woodland (total $r^2 = 0.750$, $P < 0.0001$). Stand area did not improve the model significantly.

10.3.2 Colonization patterns

Three groups of species could be distinguished according to their frequency in different stand types (Table 10.2). The first group included eight species that are equally frequent in ancient stands and in both types of older plantations. Most of these species have also colonized some of the young plantations. The second group included 13 species that easily colonize adjacent plantations but are rare in isolated stands. The third group consisted of eight species that are less frequent in both types of older plantations. Four of these are very rare outside ancient stands (*Gagea lutea, Maianthemum bifolium, Polygonatum multiflorum* and *Anemone ranunculoides*, Table 10.2). All 19 other species were too rare for statistical analysis. Among them, however, *Epipactis helleborine* had colonized six plantations, and *Brachypodium sylvaticum* was found in five plantations.

The first group of good colonizers contained more species with adaptations for long-distance dispersal (seven species with adhesive, ingested and wind-dispersed seeds) than the third group, with poor colonizers (two species, $P = 0.0117$, chi-squared statistic).

There was a rather close relationship between frequency of source populations (occurrences in ancient stands) and of colonizing populations in adjacent plantations (Table 10.3, Fig. 10.2a). For colonization of isolated plantations,

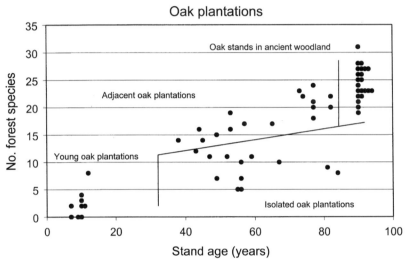

Fig. 10.1. Relationship between stand age and number of typical forest species in different types of oak plantations in the Torup-Skabersjö area, southern Sweden. Species richness of 22 oak stands in ancient woodland is given for comparison.

Table 10.3. Relationship between frequency (%) and abundance (mean cover %) of 48 forest species in different types of oak woodland in the Torup-Skabersjö area, southern Sweden.

Regression	r^2 value	P value
Frequency of species		
Ancient woodlands vs. adjacent plantations	0.752	< 0.0001
Ancient woodlands vs. isolated plantations	0.363	< 0.0001
Abundance of species		
Ancient woodlands vs. adjacent plantations	0.333	< 0.0001
as previous but excluding *Anemone nemorosa*	0.658	< 0.0001
Ancient woodlands vs. isolated plantations	0.062	0.0870
as previous but excluding *Anemone nemorosa*	0.147	0.0079

r^2-values according to linear regression analysis.

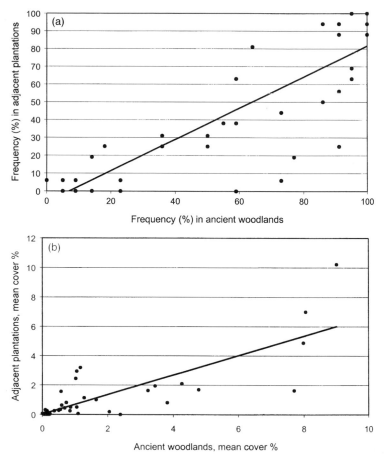

Fig. 10.2. Frequency (a) and mean cover (b, excluding *Anemone nemorosa*) of typical forest species in recent oak plantations (adjacent to ancient woodland, *n* = 16) as related to ancient oak stands (*n* = 22). Trendlines according to linear regression analysis.

source population frequency was less important, although the relationship was still highly significant.

A similar, but in general somewhat weaker, relationship appeared when analysing the role of source population abundance (mean cover %) for colonizing populations (Table 10.3, Fig. 10.2b). However, *Anemone nemorosa*, the most abundant species in ancient stands, showed a clearly deviating pattern. It has colonized most adjacent plantations (Table 10.2), but at a much lower abundance than in the ancient stands (1% mean cover in adjacent plantations, 19% mean cover in ancient oak stands).

10.3 Discussion

In oak plantations adjacent to source populations, we see a linear increase of species richness with increasing stand age. Colonization starts only a few years after stand establishment and is probably facilitated by rapid canopy closure at a stand age of *c.* 10 years, as the oaks are planted at high densities. Colonization proceeds at a rather constant rate and, at stand ages between 70 and 80 years, the stands approach a species richness comparable to that of many of the ancient woodlands. Isolation from source populations efficiently inhibits colonization of many species. After initial colonization of good dispersers, there is no further increase of species richness at stand ages > 40 years.

In four of the species that easily colonize isolated plantations, adhesive seed dispersal has been demonstrated experimentally by walking dogs through woodland or by analysing the fur of hunted wild boar and roe deer (Mrotzek *et al.*, 1999; Graae, 2000; Heinken, 2000; Heinken and Raudnitschka, 2002). In total, these authors documented adhesive dispersal for 12 species found in the present study area. Of these species, the best four colonizers (*Milium effusum*, *Circaea lutetiana*, *Poa nemoralis* and *Festuca gigantea*) were reported to be dispersed most efficiently in the above studies together with *Sanicula europaea*. However, the latter species is extremely rare in the present study area. Ferns such as *Dryopteris* spp. and *Athyrium filix-femina* are good colonizers, as their spores may be dispersed efficiently by wind but may also be transported in the fur of larger mammals. The remaining two species of the first group, *Ranunculus ficaria* and *Adoxa moschatellina*, have been reported to be good colonizers in other studies (Petersen, 1994; Bossuyt *et al.*, 1999). The mechanism of their efficient dispersal has not yet been analysed. All species of the first group are among the first colonizers of young plantations. (*A. moschatellina* was not found in oak plantations but has been recorded in young beech plantations on former fields in the study area.)

In 18 species, occasional colonization of isolated plantations has occurred, but it seems clear that the degree of stand isolation in the study area (mean distance to ancient woodland 720 m for isolated plantations) implies that colonization is severely dispersal-limited. In adjacent oak plantations, dispersal limitation seems to be of minor importance. However, four species show a much lower frequency than in ancient stands: *Anemone ranunculoides*, *Gagea lutea*, *Maianthemum bifolium* and *Polygonatum multiflorum*. It is difficult to explain the deviating behaviour of these species. For *G. lutea*, low seed production in combination with poor dispersal may result in its low colonizing capacity. In *A. ranunculoides*, most of the large source populations are found in the north-eastern part of the study area, quite far away from most plantations. *M. bifolium* is generally more frequent on rather acid soils. However, oak plantations on old fields in the present study area show a considerably higher soil pH than ancient oak stands (ten Brink, 2002) which may inhibit the colonization of acid-tolerant species such as *M. bifolium*, *Trientalis europaea* or

Gymnocarpium dryopteris. Alternatively, *M. bifolium* and *P. multiflorum* may lack suitable dispersers in the study area, especially birds that feed on their berries.

All species that are poor colonizers of isolated stands in the present study are reported as ancient woodland species elsewhere in Europe (for reviews see Wulf, 1997; Hermy *et al.*, 1999). As most of these species readily colonize adjacent plantations (e.g. Brunet and von Oheimb, 1998a; Bossuyt *et al.*, 1999), we may conclude that dispersal limitation is more important than both seed limitation and establishment limitation in the fragmented part of the study area. The relatively high base saturation in soils of old-field plantations may favour recruitment of most forest species. However, in areas with heavy eutrophication (N, P), competition from species such as *Urtica dioica* may have a negative effect (Honnay *et al.*, 1999; Verheyen and Hermy, 2001a).

My results indicate that size of source populations, which in many cases may indicate the size of the seed source, is a very important factor determining colonization rates in contiguous mixed-history forests. However, in areas with strongly fragmented forest patches, colonization rates may be largely controlled by the degree of long-distance diaspore dispersal with animals or wind. Thus, the relative importance of dispersal type increases with the degree of forest fragmentation.

In this study, I have used a rather coarse estimate of population size. The promising results may encourage more elaborate studies on diaspore production of source population and their importance for colonization patterns in secondary forests. Such studies, in combination with more experimental work, especially on epi- and endozoochoric dispersal by animals, are needed for understanding the effects of forest fragmentation on colonization patterns of plant species. The available studies indicate that many species use more than one dispersal vector. The commonly used classification of dispersal types, based on morphological characters, does not account for that, and should be completed with a more differentiated system, based on empirical and experimental data of dispersal patterns. It has been shown, for example, that *Melica uniflora*, traditionally classified as ant-dispersed, might also be dispersed quite regularly by mammals (Graae, 2000). This is because parts of the inflorescence containing seeds attach easily to fur.

10.4 Conclusions

Colonization rates within contiguous forest landscapes are largely determined by the size of local source populations. In plantations adjacent to source populations, a species-rich field layer vegetation comparable to ancient woodland can be restored within a period of 70–80 years. However, development of a carpet of spring geophytes (*Anemone* sp., *Gagea* sp., *Corydalis* sp.) may take a much longer period (cf. Brunet and von Oheimb, 1998b).

Isolated plantations are only colonized by 10–12 good dispersers among the group of forest species. In landscapes with highly fragmented forest patches, interspecific differences in the ability of long-distance dispersal by animals or wind thus seem to control colonization patterns to a large extent.

References

Bellemare, J., Motzkin, G. and Foster, D.R. (2002) Legacies of the agricultural past in the forested present: an assessment of historical land-use effects on rich mesic forests. *Journal of Biogeography* 29, 1401–1420.

Bossuyt, B., Hermy, M. and Deckers, J. (1999) Migration of herbaceous plant species across ancient–recent forest ecotones in central Belgium. *Journal of Ecology* 87, 628–638.

Brunet, J. and von Oheimb, G. (1998a) Migration of vascular plants to secondary woodlands in southern Sweden. *Journal of Ecology* 86, 429–438.

Brunet, J. and von Oheimb, G. (1998b) Colonization of secondary woodlands by *Anemone nemorosa*. *Nordic Journal of Botany* 18, 369–377.

Butaye, J., Jaquemyn, H. and Hermy, M. (2001) Differential colonization causing non-random forest plant community structure in a fragmented agricultural landscape. *Ecography* 24, 369–380.

Dzwonko, Z. (1993) Relation between the floristic composition of isolated young woods and their proximity to ancient woodland. *Journal of Vegetation Science* 4, 693–698.

Dzwonko, Z. (2001) Effect of proximity to ancient deciduous woodland on restoration of the field layer vegetation in a pine plantation. *Ecography* 24, 198–204.

Dzwonko, Z. and Loster, S. (1992) Species richness and seed dispersal to secondary woods in southern Poland. *Journal of Biogeography* 19, 195–204.

Graae, B.J. (2000) The impact of forest continuity on the flora in Danish deciduous forests. PhD thesis, University of Copenhagen, Denmark.

Graae, B.J. and Sunde, P.B. (2000) The impact of forest continuity and management on forest floor vegetation evaluated by species traits. *Ecography* 23, 720–731.

Grashof-Bokdam, C.J. and Geertsema, W. (1998) The effect of isolation and history on colonization patterns of plant species in secondary woodland. *Journal of Biogeography* 25, 837–846.

Heinken, T. (2000) Dispersal of plants by a dog in a deciduous forest. *Botanische Jahrbücher Systematik* 122, 449–467.

Heinken, T. and Raudnitschka, D. (2002) Do wild ungulates contribute to the dispersal of vascular plants in central European forests by epizoochory? A case study in NE Germany. *Forstwissenschaftliches Centralblatt* 121, 179–194.

Hermy, M., Honnay, O., Firbank, L., Bokdam-Grashof, C. and Lawesson, J.E. (1999) An ecological comparison between ancient and other forest plant species of Europe, and the implication for forest conservation. *Biological Conservation* 91, 9–22.

Honnay, O., Degroote, B. and Hermy, M. (1998) Ancient-forest plant species in western Belgium: a species list and possible ecological mechanisms. *Belgian Journal of Botany* 130, 139–154.

Honnay, O., Hermy, M. and Coppin, P. (1999) Impact of habitat quality on forest plant species colonization. *Forest Ecology and Management* 115, 157–170.

Jordano, P. and Godoy, J.A. (2002) Frugivore generated seed shadows: a landscape view of demographic and genetic effects. In: Levey, D.J., Silva, W.R. and Galetti, M. (eds) *Seed Dispersal and Frugivory: Ecology, Evolution and Conservation.* CAB International, Wallingford, UK, pp. 305–321.

Matlack, G.R. (1994) Plant species migration in a mixed-history forest landscape in eastern North America. *Ecology* 75, 1491–1502.

Mrotzek, R., Halder, M. and Schmidt, W. (1999) Die Bedeutung von Wildschweinen für die Diasporenausbreitung von Phanerogamen. *Verhandlungen Gesellschaft für Ökologie* 29, 437–443.

Peterken, G.F. and Game, M. (1984) Historical factors affecting the number and distribution of vascular plant species in woodlands of central Lincolnshire. *Journal of Ecology* 72, 155–182.

Petersen, P.M. (1994) Flora, vegetation, and soil in broadleaved ancient and planted woodland, and shrub on Røsnæs, Denmark. *Nordic Journal of Botany* 14, 693–709.

Singleton, R., Gardescu, S., Marks, P.L. and Geber, M.A. (2001) Forest herb colonization of postagricultural forests in central New York State, USA. *Journal of Ecology* 89, 325–338.

ten Brink, D.-J. (2002) Land use history affects present day forest soil properties. BSc thesis, Department of Ecology, Lund University, Sweden.

Thompson, K., Bakker, J.P., Bekker, R.M. and Hodgson, J.G. (1998) Ecological correlates of seed persistence in soil in the north-west European flora. *Journal of Ecology* 86, 163–169.

Verheyen, K. and Hermy, M. (2001a) An integrated analysis of the spatial-temporal colonization patterns of forest plant species. *Journal of Vegetation Science* 12, 567–578.

Verheyen, K. and Hermy, M. (2001b) The relative importance of dispersal limitation of vascular plants in secondary forest succession in Muizen Forest, Belgium. *Journal of Ecology* 89, 829–840.

Weimarck, H. and Weimarck, G. (1985) *Atlas över Skånes Flora.* Förlagstjänsten, Stockholm.

Whitney, G.G. and Foster, D.R. (1988) Overstorey composition and age as determinants of the understorey flora of woods of central New England. *Journal of Ecology* 76, 867–876.

Wittig, R., Gödde, R., Neite, H., Papajewski, W. and Schall, O. (1985) Die Buchenwälder auf den Rekultivierungsflächen im Rheinischen Braunkohlerevier: Artenkombination, pflanzensoziologische Stellung und Folgerungen für zukünftige Rekultivierungen. *Angewandte Botanik* 59, 95–112.

Wulf, M. (1997) Plant species as indicators of ancient woodland in northwestern Germany. *Journal of Vegetation Science* 8, 635–642.

Zar, J.H. (1996) *Biostatistical Analysis*, 3rd edn. Prentice-Hall International, London.

Multiple-scale Factors Affecting the Development of Biodiversity in UK Plantations

11

J.W. Humphrey,[1] A.J. Peace,[2] M.R. Jukes[2] and E.L. Poulsom[1]

[1]Forest Research, Northern Research Station, Roslin, Midlothian, UK;
[2]Forest Research, Alice Holt Research Station, Wrecclesham, Surrey, UK

The effects of climate, site quality, stand structure and landscape on the development of biodiversity within plantations were investigated, with particular emphasis on woodland vascular plants, bryophytes, mycorrhizal fungi, and subcanopy Coleoptera. Data were obtained as part of a broad study of biodiversity within Scots pine (*Pinus sylvestris*), Corsican pine (*Pinus nigra* var. *maritima*), Sitka spruce (*Picea sitchensis*) and Norway spruce (*Picea abies*) plantations of different ages and in different climate zones (uplands, foothills, lowlands) across Britain. There was a significant effect of climate zone on the species-richness of all four taxonomic groups, with upland stands having the highest bryophyte counts and lowland stands having the highest Coleoptera counts. Species counts of woodland vascular plants and mycorrhizals were negatively correlated with distance to nearest ancient semi-natural woodland. Counts of bryophytes were positively correlated with the amount of semi-natural woodland within 1 km of the stand, and Coleoptera counts were positively correlated with the amount of non-wooded semi-natural vegetation within 1 km. Other factors such as tree species, soil and stand structure were significantly related to counts of woodland vascular plants and bryophytes. Stand age and land-use history did not feature in the best-fit regression models. The development of biodiversity in plantations is clearly influenced by the proximity and amount of semi-natural vegetation within the landscape, suggesting that to be cost-effective, management for restoration and enhancement of biodiversity in planted stands should be targeted towards those stands in 'semi-natural landscapes' where agricultural land-use has historically been of low intensity.

©CAB International 2004. *Forest Biodiversity: Lessons from History for Conservation*
(eds O. Honnay, K. Verheyen, B. Bossuyt and M. Hermy)

11.1 Introduction

Like many other countries in Europe, Britain has a long history of clearance
and fragmentation of natural forest. By the beginning of the 20th century
only 4–5% of the land area was occupied by ancient semi-natural woodland
(ASNW), the modified remnants of the original 'wildwood' (Peterken, 1996).
Woodland biodiversity was severely affected by the loss and fragmentation of
forest habitats, and species such as capercaillie (*Tetrao urogallus*) and others
were driven (sometimes temporarily) to extinction. During the 20th century
there was extensive commercial afforestation of much of the upland area
of Britain, using exotic conifer species such as Sitka spruce [*Picea sitchensis*
(Bong.) Carr.], Norway spruce [*P. abies* (L.) Karst.] and native conifer species
such as Scots pine (*Pinus sylvestris* L.). Lowland areas were planted with
species such as Corsican pine (*P. nigra* var. *maritima* Arnold). About two-thirds
of new planting was on open land (Kirby, 1993), the remainder within existing
semi-natural woodland (Roberts *et al.*, 1992; Spencer and Kirby, 1992). The
present forest cover in Britain is around 11.5% of total land area (Forestry
Commission, 2002).

Current government policy and associated incentives promote the
restoration of wooded landscapes (Anon., 1999, 2000, 2002; Humphrey *et al.*,
2003a) through the publication of action plans for priority woodland habitats
(Anon., 1995). These plans contain area targets for the rehabilitation of
degraded ASNW; the conversion of planted ancient woodland sites (PAWS)
back to semi-natural woodland, and the creation of new native woodlands. Complementing this habitat restoration is action for key species of
conservation concern. Plans have been produced for over 380 species, many
of which use woodland habitats (Anon., 1995).

Recently, attention has been given to the positive role that plantations
could play in the restoration of woodland habitats and biodiversity (Humphrey
et al., 2001; Hartley, 2002; Honnay *et al.*, 2002). Given the current economic
downturn in timber prices, and continued high recreational use of forests
(Forestry Commission, 2002), there is considerable interest in improving
plantations for biodiversity and other non-timber benefits, particularly in the
economically marginal upland areas. However, despite some recent studies
(e.g. Petty *et al.*, 1995), little is known about the relative influence of different
factors on the development of biodiversity in planted forests. Structural
features such as deadwood, canopy tree species and vertical foliage cover have
been shown to correlate well with the species richness and composition of
invertebrate (Humphrey *et al.*, 1999; Jukes *et al.*, 2001, 2002), fungal and
lower plant communities (Humphrey *et al.*, 2000, 2002a). Plant communities
are also influenced by soil and climatic factors, as well as stand structure (Ferris
et al., 2000).

Research elsewhere has shown the importance of land-use history and
other landscape-scale factors (e.g. proximity of sources of colonizing species) in

influencing the development of biodiversity within secondary woodland (Grashof-Bokdam and Geertsema, 1998; Hilmo and Såstad, 2001; Verheyen and Hermy, 2001; Honnay *et al.*, 2002). The development of digital datasets of ancient and secondary woodland cover in Britain provided an opportunity to evaluate the effects of site history and landscape-scale factors on plantation biodiversity across Britain, using existing floral and faunal data obtained from the range of stands described in Humphrey *et al.* (2001). We tested the hypothesis that proximity to semi-natural woodland and site history (whether ancient woodland site or not) had a greater influence on biodiversity development in the planted stands than stand structure, soils and climate. Preliminary results are presented for woodland vascular plants, bryophytes, mycorrhizal fungi and subcanopy Coleoptera.

11.2 Materials and Methods

11.2.1 Study sites

Study sites were established in plantations within three climate zones (Table 11.1). These zones were defined by annual precipitation totals (Pyatt *et al.*, 2001): 'lowlands' < 800 mm; 'foothills' 800–1500 mm and 'uplands' > 1500 mm. At the majority of sites, a chronosequence of four 1 ha (100 m × 100 m) permanent sample plots was established in forest stands reflecting different growth stages based on the normal economic rotation. These were:

1. Pre-thicket – restock sites, crop height 2–4 m, age 8–10 years;
2. Mid-rotation – crop height 10–20 m, age 20–30 years;
3. Mature – crop height 20–25 m, age 50–80 years; and
4. Overmature (in all forest types excluding Norway spruce and foothills Scots pine) – crop height 25+ m, age 68–80 years.

11.2.2 Site history and landscape-scale factors

Information on the land-use history of the plots was obtained from the national inventory of ancient woodland (Roberts *et al.*, 1992; Spencer and Kirby, 1992) and from pre-planting records held within Forestry Commission archives. The ancient woodland inventory was compiled during the 1980s and 1990s to provide baseline information on the distribution, amount and condition of ancient woodland across Britain. Ancient woodlands are those known to have existed since at least AD 1600 in England and Wales, and 1750 in Scotland (Roberts *et al.*, 1992; Spencer and Kirby, 1992). Ancient woodlands are judged to have higher nature conservation value than woods of more recent origin. They have been further categorized by stand condition; either semi-natural,

Table 11.1. Location and details of assessment plots. Digits after the decimal place in the site number indicate: 1, pre-thicket; 2, mid-rotation; 3, mature; 4, overmature.

Site number	Site	Longitude/latitude	Climatic zone	Crop species	Age (years)	Previous land use
1	**Glen Affric**					
1.1	Lochan Dubh, Cannich	57°23'N 4°48'W	Foothills	Scots pine	12	Heath/grassland
1.2	Knockfin	57°18'N 4°51'W	Foothills	Scots pine	35	Heath/grassland
1.3	Plodda Falls	57°17'N 4°55'W	Foothills	Scots pine	96	Heath/grassland
2	**Strathspey**					
2.1	Moor of Alvie	57°9'N 3°54'W	Foothills	Scots pine	8	Heath/grassland
2.2	An Slugan	57°12'N 3°45'W	Foothills	Scots pine	32	Ancient woodland
2.3	Glenmore Lodge	57°10'N 3°40'W	Foothills	Scots pine	64	Ancient woodland
3	**Thetford**					
3.1	Lynford	52°29'N 0°42'E	Lowlands	Scots pine	18	Heath/grassland
3.2	Horsford Woods	52°43'N 1°15'E	Lowlands	Scots pine	37	Heath/grassland
3.3	High Lodge	52°26'N 0°41'E	Lowlands	Scots pine	68	Heath/grassland
4	**New Forest/Windsor**					
4.1	Knightwood Inclosure	50°51'N 1°38'W	Lowlands	Scots pine	26	Heath/grassland
4.2	Denny Lodge	50°51'N 1°32'W	Lowlands	Scots pine	49	Heath/grassland
4.3	Denny Lodge	50°50'N 1°31'W	Lowlands	Scots pine	66	Heath/grassland
4.4	The Look Out, Windsor	51°23'N 0°44'W	Lowlands	Scots pine	66	Heath/grassland
5	**Knapdale**					
5.1	Dunardy	56°4'N 5°31'W	Uplands	Sitka spruce	9	Heath/grassland
5.2	Dunardy	56°4'N 5°31'W	Uplands	Sitka spruce	24	Heath/grassland
5.3	Kilmichael	56°5'N 5°19'W	Uplands	Sitka spruce	44	Heath/grassland
5.4	Dunardy	56°4'N 5°30'W	Uplands	Sitka spruce	62	Heath/grassland
6	**Clunes**					
6.1	South Laggan	57°0'N 4°52'W	Uplands	Sitka spruce	8	Heath
6.2	Clunes	56°58'N 4°59'W	Uplands	Sitka spruce	28	Ancient woodland

6.3	Clunes	56°58'N 4°59'W	Uplands	Sitka spruce	62	Ancient woodland
6.4	South Laggan	57°0'N 4°53'W	Uplands	Sitka spruce	67	Heath/grass
7	**Forest of Dean**					
7.1	Ruddle Marsh	51°49'N 2°34'W	Lowlands	Norway spruce	15	Ancient woodland
7.2	Cannop	51°48'N 2°34'W	Lowlands	Norway spruce	31	Ancient woodland
7.3	Ruardean	51°51'N 2°32'W	Lowlands	Norway spruce	56	Ancient woodland
8	**Fineshade**					
8.1	Fineshade	52°34'N 0°33'W	Lowlands	Norway spruce	17	Ancient woodland
8.2	Fineshade	52°34'N 0°34'W	Lowlands	Norway spruce	39	Ancient woodland
8.3	Fermyn Woods	52°27'N 0°33'W	Lowlands	Norway spruce	65	Grassland
9	**Kielder**					
9.1	Falstone	55°10'N 2°27'W	Foothills	Sitka spruce	6	Grassland
9.2	Falstone	55°9'N 2°27'W	Foothills	Sitka spruce	23	Grassland
9.3	Falstone	55°9'N 2°31'W	Foothills	Sitka spruce	57	Grassland
9.4	Archie's Rigg	55°8'N 2°28'W	Foothills	Sitka spruce	69	Grassland/mire
10	**Glentress**					
10.1	Glentress	55°40'N 3°9'W	Foothills	Sitka spruce	10	Heathland
10.2	Glentress	55°39'N 3°8'W	Foothills	Sitka spruce	28	Heath/grassland
10.3	Glentress	55°40'N 3°9'W	Foothills	Sitka spruce	55	Heath/grassland
10.4	Cardrona	55°37'N 3°6'W	Foothills	Sitka spruce	61	Grassland
11	**Thetford**					
11.1	Kings Forest	52°21'N 0°40'E	Lowlands	Corsican pine	8	Heath/grassland
11.2	Kings Forest	52°21'N 0°40'E	Lowlands	Corsican pine	33	Heath/grassland
11.3	Kings Forest	52°20'N 0°39'E	Lowlands	Corsican pine	59	Heath/grassland
11.4	High Lodge	52°26'N 0°40'E	Lowlands	Corsican pine	69	Heath/grassland
12	**Clipstone**					
12.1	Clipstone	53°9'N 1°6'W	Lowlands	Corsican pine	9	Heath/grassland
12.2	Clipstone	53°9'N 1°3'W	Lowlands	Corsican pine	43	Heath/grassland
12.3	Clipstone	53°10'N 1°5'W	Lowlands	Corsican pine	49	Heath/grassland

consisting of trees and shrubs arising from natural regeneration or coppice regrowth, or planted, comprising stands of native or non-native conifer and broadleaved tree species (Roberts et al., 1992; Spencer and Kirby, 1992). The lower size limit for inclusion in the inventory is 2 ha.

Information on the distribution of ancient woodland (held as a data layer within ArcView GIS) was used to calculate the distance from the centre of each assessment plot to the centre of the nearest ASNW ('nearest' was defined as the ASNW polygon with the nearest edge to the plot), and the percentage cover of semi-natural woodland within a 1 km 'buffer' surrounding the plot. Buffer-type measures are considered to give more robust indicators of connectivity than simple nearest-neighbour estimates (Moilanen and Nieminen, 2002). The percentage cover of non-wooded (open) semi-natural vegetation within 1 km of each was estimated from site maps. The limit of 1 km was used as this represents a rough measure of maximum dispersal distance averaged over a range of different woodland species (Humphrey et al., 2003b).

11.2.3 Stand structure and mensuration assessments

Assessment methods for the full range of species groups, stand structure and soil variables recorded in the plots are summarized in Humphrey et al. (2001) and described in more detail in Humphrey et al. (2003c). Table 11.2 lists the categorical variables used in the analysis, Table 11.3 lists the other parameters.

Assessments of diameter at breast height (DBH), height to the base of the live crown, and top height, were made for all living trees ≥ 7 cm DBH in 100 m^2 or 625 m^2 sub-plots (depending on tree density) within each 1 ha plot (see Humphrey et al., 2003d for details). Basal area was calculated for each 1 ha plot, following Hamilton (1998). Vertical stand structure was assessed using a visual cover method. Four vegetation strata were defined: S_1 (field) 10 cm to 1.9 m in height; S_2 (shrub) 2–5 m; S_3 (lower canopy) 5.1–15 m; and S_4 (upper canopy) 15.1–20 m. Percentage cover of vegetation within each vertical stratum was described to the nearest 5% and expressed as a mean of 16 measures taken from each 1 ha plot. To convert these cover values to a unified

Table 11.2. Categorical data used in the analysis. Values in parentheses are number of plots.

Variable	Description	Values
AW	Plot on ancient woodland site	Yes (9); no (33)
SPS	Main tree species	Pine (20); spruce (22)
REGION	Climate zone	Uplands (8); foothills (14); lowlands (20)
CHRONO	Chronosequence age	Pre-thicket (12); mid-rotation (12); mature (12); overmature (6)

measure of stand structure, a cover index (SINDEX) was calculated using the formula:

$$SINDEX = 1.9s_1 + 3s_2 + 10s_3 + 5s_4$$

Table 11.3. Landscape-setting, stand structure, and species variables used in the analysis.

Variable	Description	Mean	Max.	Min.	Standard deviation
DASNW	Distance of plot to nearest ASNW (m)	2582	11349	0	3027
PCSNW	% cover of semi-natural woodland within 1 km	11	68	0	13
PCOSN	% cover of open semi-natural vegetation within 1 km	10	50	0	13
PCA1	1st principal component of soil variables ($\uparrow$P, K, Mg, organic matter and NH_4^+)	0	1.7	−1.0	0.6
PCA2	2nd principal component of soil variables ($\uparrow$pH, Ca and NO_3^-, $\downarrow$organic matter and NH_4^+)	0	1.6	−0.7	0.5
PCA3	3rd principal component of soil variables ($\uparrow$Mg, $\downarrow NH_4^+$ and NO_3^-)	0	0.9	−0.9	0.4
PCA4	4th principal component of soil variables ($\uparrow NO_3^-$, $\downarrow$pH)	0	0.8	−0.7	0.3
S_1	Vertical structure in the field layer (% cover)	20	73	0	22
S_2	Vertical structure in the shrub layer (% cover)	6	41	0	10
S_3	Vertical structure in the subcanopy (% cover)	14	52	0	16
S_4	Vertical structure in the upper canopy (% cover)	8	32	0	10
SINDEX	Vertical structure index	243	652	47	166
TOPHT	Top height (m)	6	12	1	3
HTLC	Height to live crown (m)	3	8	0	2
MBA	Measurable basal area (m²/ha)	29	60	0	17
DEADF	Volume of fresh deadwood (m³/ha)	23	285	0	48
DEADR	Volume of rotten deadwood (m³/ha)	13	87	0	17
LITTER	Depth of litter (cm)	3	8	0	2.1
LAI	Leaf area index	3	10.2	0.5	2.1
CWVP	Number of woodland vascular plant species	4	8	0	2.4
CCOL	Number of malaise-trapped Coleoptera species	28	57	12	12
CBRY	Number of bryophyte species	13	23	3	5
CMYCOR	Number of mycorrhizal species	16	53	0	13

where s_1–s_4 are the values for field, shrub, lower canopy and upper canopy strata, and numbers refer to the depth of each stratum in metres.

The leaf area index (LAI) for each plot was calculated using the canopy radiation model of Goudriaan (1988) for diffuse light conditions. Data for the model came from measurements of photosynthetically active radiation, obtained using a hand-held sunfleck ceptometer (Decagon Instruments, Pullman, Washington, USA).

Accumulations of deadwood were recorded along two transects bisecting the 1 ha plot. Total volume of fallen deadwood ($\geq$ 7 cm DBH) was calculated using the line intercept method (Kirby *et al.*, 1998). Standing deadwood was measured within 5 m either side of the transect. Volumes of fallen and standing deadwood were calculated by assuming that all pieces were cylindrical. Deadwood quality was described using a visual five-point scale, where categories 1–3 corresponded to 'fresh' decay classes (DEADF), 4 and 5 to advanced decay (DEADR) (*sensu* Hunter, 1990).

11.2.4 Soil and vegetation

Soil samples were taken from a range of depth and locations within each 1 ha plot and bulked to give one sample per plot. Details of the chemical analyses carried out are given in Ferris *et al.* (2000). Many of the soil chemistry variables were inter-correlated, so summative vectors/gradients were calculated using principal components analysis (Table 11.3). Field- and ground-layer vegetation composition was assessed visually within 2 × 2 m quadrats centred in the eight subplots described in Section 11.2.3. The DOMIN scale was used to assess cover abundance (see Ferris *et al.*, 2000). Litter depth to the nearest centimetre was also assessed in these subplots. The vegetation data were divided into two components: bryophytes and woodland vascular plants. The bryophyte species count was augmented by adding records of species found on deadwood; see Humphrey *et al.* (2002a) for details of survey methodology. Woodland vascular plants were defined as being preferentially associated with woodland habitats (using Rodwell, 1991) and having a measure of shade tolerance (defined as having a Hill–Ellenberg L-value of 6 or less (Hill *et al.*, 1999). These species cannot be considered as 'ancient woodland indicators' in the strict sense (Peterken, 1993). Total species counts per 1 ha plot were used in the analysis.

11.2.5 Mycorrhizals

The presence/absence of mycorrhizal sporocarps was recorded in the eight subplots described in Section 11.2.3 (see Humphrey *et al.*, 2000 for details). Assessments were made over the August–October period to coincide with the main time of sporocarp production. Three visits were made to each plot at roughly monthly intervals over this period, and repeated over 3 or 4 years. As with the vascular plants and bryophytes, total mycorrhizal species counts per 1 ha plot were used in the analysis.

11.2.6 Subcanopy coleopterans

Invertebrates in the subcanopy space in each plot were sampled by a single Malaise trap, operating continuously for a 20-week period for 2 consecutive years. For details of the methodology, see Humphrey *et al.* (1999). Species counts for Coleoptera were obtained from cumulative catches over this 2-year period.

11.2.7 Analysis

The effects of the environmental parameters (Tables 11.2 and 11.3) on the species count data were analysed using generalized linear models (GLMs) with a standard log link function and Poisson error distribution (McCullagh and Nelder, 1989). Significant variables were selected by stepwise regression to produce 'best-fit' models. These models included the combination of parameters which best predicted the response of the species count data (i.e. where deviance was minimized). Separate models were generated for each species group. Approximate *F*-tests were used to estimate the significance of the regression models.

11.3 Results and Discussion

11.3.1 General linear models

In the summary GLM outputs (Table 11.4) the value of the residual mean deviance gives an indication of how well each model accounts for the 'noise' in the species count data. The closer the value of the mean deviance is to one, the more noise is accounted for. On this basis, the models for woodland vascular plants and bryophytes appear to give the 'best fit', with the mycorrhizal model giving the poorest fit. However, all four regression analyses were significant at the 0.1% level (Table 11.4). Estimates of the main parameters contributing to the best fit models are given in Table 11.5. The sign for these parameters indicates whether the parameter was negatively or positively related to the species counts.

Table 11.4. Summary of multiple linear regressions of species counts on the set of environmental variables listed in Tables 11.2 and 11.3: (a) woodland vascular plants (CWVP); (b) bryophytes (CBRY); (c) mycorrhizals (CMYCOR); (d) subcanopy Coleoptera (CCOL).

	Degrees of freedom	Deviance	Mean deviance	Deviance ratio	Significance (*P*) of approximate *F* values
(a) CWVP					
Regression	5	40.11	8.0230	9.27	< 0.001
Residual	36	31.15	0.8652		
Total	41	71.26	1.7381		
(b) CBRY					
Regression	3	29.33	9.778	8.84	< 0.001
Residual	38	42.04	1.106		
Total	41	71.37	1.741		
(c) CMYCOR					
Regression	6	286.6	47.765	9.99	< 0.001
Residual	34	162.6	4.784		
Total	40	449.2	11.231		
(d) CCOL					
Regression	3	128.19	42.731	24.73	< 0.001
Residual	38	65.67	1.728		
Total	41	193.86	4.728		

11.3.2 Climatic influences

For all four species groups, climatic zone had a significant influence on the distribution of species counts (Table 11.5). In the case of woodland vascular plants, bryophytes and mycorrhizals, plots in the uplands climatic zone had a higher proportion of species counts than plots in the lowlands and foothills, whereas the opposite was true for the subcanopy Coleoptera, where lowland plots were more species-rich (Table 11.5). For both bryophytes and Coleoptera, climate zone was the key factor affecting species counts (Fig. 11.1). These results confirm earlier conclusions (Humphrey *et al.*, 2002a; Jukes *et al.*, 2002) that climate zone has a key influence on species-richness of bryophytes and invertebrates, with bryophytes more abundant in the cool, wet climate of north and west Britain and invertebrates associated with drier conditions in the south of the country.

11.3.3 Effects of landscape-scale factors

There were significant effects of landscape-scale factors on all species groups (Table 11.5). For woodland vascular plants, distance to nearest ASNW was

Table 11.5. Estimates of parameters from regressions of species counts on environmental variables: (a) woodland vascular plants (CWVP); (b) bryophytes (CBRY); (c) mycorrhizals (CMYCOR); (d) subcanopy Coleoptera (CCOL). The key to variables is given in Tables 11.2 and 11.3. *t*-Tests were used to determine whether the estimates of parameter values were significantly greater or less than zero. The antilog of the estimates indicates the proportion of change in species counts in relation to the change in parameter values.

Variable	Parameter estimate	Standard error	t	Significance (P) of t values	Antilog of estimate
(a) CWVP					
Constant	1.731	0.174	9.92	< 0.001	5.644
DASNW	−0.0001510	0.0000370	−4.08	< 0.001	0.9998
PCA1	−0.331	0.134	−2.47	0.018	0.7180
LITTER	−0.1502	0.0448	−3.35	0.002	0.8605
REGION – lowlands	0.414	0.210	1.97	0.057	1.513
REGION – uplands	0.543	0.213	2.55	0.015	1.722
(b) CBRY					
Constant	2.853	0.0679	41.99	< 0.01	17.34
REGION – lowlands	−0.148	0.0965	−1.53	0.133	0.8624
REGION – uplands	0.385	0.0991	3.89	0.001	1.47
PCSNW	0.0064	0.00262	2.53	0.015	1.007
(c) CMYCOR					
Constant	2.948	0.190	15.48	< 0.001	19.07
DASNW	−0.0001224	0.0000435	−2.82	0.008	0.99
PCA1	0.521	0.141	3.68	< 0.001	1.683
S2	0.0247	0.00954	2.59	0.014	1.025
SPS – spruce	−0.798	0.257	−3.10	0.004	0.4503
REGION – lowlands	0.069	0.239	0.29	0.776	1.071
REGION – uplands	0.710	0.288	2.46	0.019	2.034
(d) CCOL					
Constant	2.911	0.0825	35.29	< 0.001	18.38
REGION – lowlands	0.631	0.0932	6.77	< 0.001	1.880
REGION – uplands	−0.101	0.137	−0.74	0.465	0.9037
PCOSN	0.008	0.00282	2.84	0.007	1.008

negatively related to the modelled species counts when the effects of other factors were held constant (Fig. 11.2). This suggests that stands nearer to ancient woodland are more likely to acquire woodland species and implies that the development of woodland plant communities in these secondary woods may be dispersal limited, as suggested by Hermy *et al.* (1999). The likelihood of a stand containing woodland species drops off significantly over distances of 2000 m or more from the centre of the putative source woodland (Fig. 11.2). This distance is significantly greater than maximal dispersal distances (over periods of one or more centuries) that have been recorded for

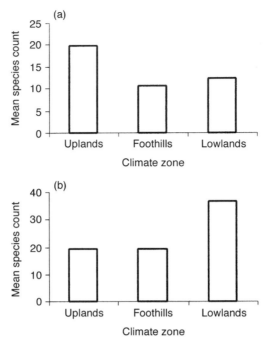

Fig. 11.1.   Observed mean species counts for: (a) bryophytes and (b) sub-canopy Coleoptera in relation to climate zone.

woodland plants by Brunet and von Oheimb (1998), Grashof-Bokdam and Geertsema (1998) and Vickers *et al.* (2000). This apparent difference may be due to the occurrence, in our study, of woodland species surviving in small patches within the landscape. These would not have been picked up in the analysis, because only woods of 2 ha or more are recorded in the ancient woodland inventory. Also, this current analysis did not take into account the dispersal mechanisms of the woodland species present in the plantations, so it is possible that only species with superior powers of dispersal (e.g. wind dispersers) were present in the more isolated plantations.

There was no effect of stand age on the species richness of woodland vascular plants. The chronosequence parameter did not figure significantly in the optimal regression models. Humphrey *et al.* (2003d), drawing on a larger plant dataset than that used in our study, found that the character of the vegetation became progressively more 'wooded' across the chronosequence. The difference between the two studies lies in the definition of woodland vegetation. Humphrey *et al.* (2003d) used a much wider pool of species, including bryophytes, lichens and woodland-glade species, and related both species occurrence and abundance to summative values for National Vegetation Woodland Communities (Rodwell, 1991).

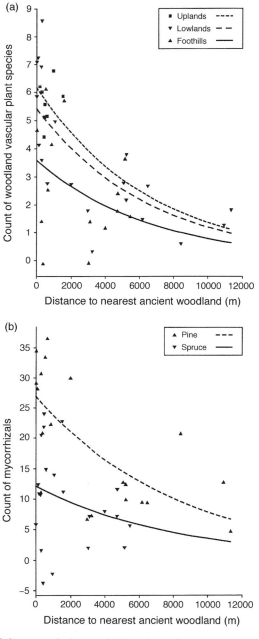

Fig. 11.2. Fitted (lines) and observed (triangles) relationships between landscape-scale factors and species counts: (a) woodland vascular plants in relation to distance from ancient semi-natural woodland (ASNW); (b) mycorrhizals in relation to distance from ASNW.

Continued

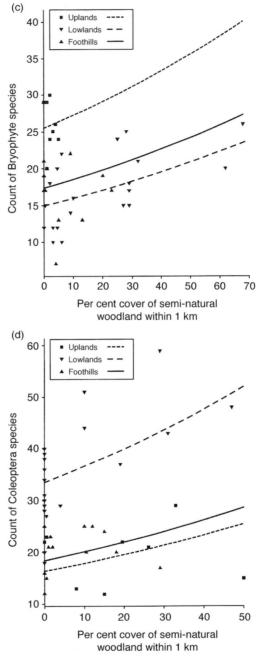

Fig. 11.2. *Continued.* Fitted (lines) and observed (triangles) relationships between landscape-scale factors and species counts: (c) bryophytes in relation to percentage cover of semi-natural woodland within 1 km; (d) subcanopy Coleoptera in relation to percentage cover of open semi-natural vegetation within 1 km.

Little research has been done on factors affecting mycorrhizal communities within planted stands (Humphrey *et al.*, 2000). Many of the species recorded in our study are rare and threatened, and previously thought to be restricted to ancient native Scots pine woodland (Humphrey *et al.*, 2000). The data suggest (Fig. 11.2b) that successful dispersal over considerable distances to new habitats is possible if habitat conditions are right. Judging by the influence of stand structure (Table 11.5), mycorrhizals appear to be most species-rich in stands with dense foliage cover in the shrub layer (s_2). In most cases this cover is made up of the lower branches of the planted crop, creating a dark and moist microclimate. These conditions restrict the development of competing vegetation and provide a high density of roots for potential mycorrhizal associations (Humphrey *et al.*, 2000).

While distance to source of propagules is clearly an important factor in the development of vascular plant and bryophyte communities in planted stands, the amount and type of surrounding semi-natural vegetation is also significant for bryophytes and subcanopy Coleoptera (Fig. 11.2c,d). This factor has not been considered explicitly in previous studies of landscape factors on biodiversity development (although see the recent study by Hersperger and Forman, 2003). The majority of studies have focused on woodland patches within intensive agricultural landscapes, where the matrix is relatively inhospitable to woodland plants, and distances between source and sink populations are readily quantifiable (Honnay *et al.*, 2002). In landscapes where the matrix is less intensively managed, woodland species are rarely restricted to ancient woodland and can often survive in other semi-natural habitats, such as recent native woodland, old hedges, meadows or under bracken or tall heather (Peterken, 1993). We suspect that this is the reason for the positive correlations observed between semi-natural vegetation and bryophyte and Coleoptera species-richness.

11.3.4 Site history

Surprisingly, site history did not contribute significantly to the best-fit regression models. However, it would be inadvisable on the strength of these models to extrapolate more generally about the lack of importance of site history. Recent surveys (Spencer, 2002) have shown that plantations on ancient woodland sites (PAWS) frequently contain flora and fauna associated with vestiges of semi-natural woodland habitat not eradicated by the plantation. There is considerable variability in the amount of residual habitat found on different PAWS, and we suspect that with only nine examples of PAWS in our dataset, the amount of replication was not sufficient to cover a range of situations where vestigial habitat was significant.

11.3.5 Soils

Counts of woodland vascular plants were negatively related to the first axis of the ordination of soil variables (PCA1); counts of mycorrhizal fungi were positively related (Table 11.5). PCA1 is essentially a soil nutrient gradient positively correlated with increasing abundance of magnesium, phosphorus, potassium, organic matter and ammonium. High soil nutrient status has been cited as a barrier to the recruitment of woodland vascular plants, as these are out-competed by more vigorous, ruderal species which respond to high fertility (Honnay *et al.*, 2002). Our data tend to support this contention for the woodland vascular plants, where species counts were lower on the richer sites.

11.4 Conclusions and Implications for Management

The original hypothesis that landscape-scale factors and site history would have a greater influence on biodiversity development than soil, stand and climatic factors is only supported by results for the vascular plants (and then only for landscape factors and not site history). Although landscape factors appear to be important in determining the development of biodiversity within plantations (specifically the proximity and amount of semi-natural wooded and non-wooded habitat), the effects are taxon-specific; soil, climate and stand factors all have a significant influence on species richness and need to be considered in addition.

A surprising feature of this study (and its predecessors – see Humphrey *et al.*, 2001) was the high levels of native species diversity recorded in, for the most part, planted stands of non-native conifers. Traditionally, closely planted, uniform conifer plantations have been viewed as hostile to biodiversity (Nature Conservancy Council, 1986), but in this current study, there were few significant negative effects on species counts of what might be considered the deleterious plantation factors, such as low light levels, dense canopy cover, presence of exotic crop species, etc. In fact, mycorrhizal species richness was positively correlated with dense lower canopy conditions. For most taxa, species richness in the plantations was similar to that found in semi-natural woodland (Humphrey *et al.*, 2003c).

Given that planted stands can make a positive contribution to the conservation of woodland biodiversity in Britain, it is necessary to consider how that contribution can be enhanced. Guidance has been given for specific groups and habitat features (e.g. Currie and Elliot, 1997; Humphrey *et al.*, 2000, 2002b), but this current study has emphasized the importance of landscape-scale factors, such as the proximity of wooded and non-wooded semi-natural habitat, in influencing the successful development of biodiversity within plantations. We propose therefore, that plantations within landscapes that have a high proportion of semi-natural habitat and have a history of low-intensity agricultural management should be prioritized for enhancement activities. This is being recognized increasingly in local and regional action plans for biodiversity (e.g. Cosgrove, 2002) where the habitat value of plantations, especially those that are long-established and have a diverse stand structure, is often comparable to that of the ancient woodland stands in the surrounding landscape.

Acknowledgements

We are grateful to Christine Brown of Woodland Surveys Branch in Forest Research for GIS analysis of the ancient woodland inventory. Chris Quine and Jason Hubert made helpful comments on earlier drafts. The research was funded by the Forestry Commission.

References

Anon (1995) *Biodiversity: the UK Steering Group Report*, Vol 2. Action Plans, London.
Anon (1999) *England Forestry Strategy – a New Focus for England's Woodlands*. Forestry Commission, Cambridge, UK.
Anon (2000) *Forests for Scotland: the Scottish Forestry Strategy*. Forestry Commission, Edinburgh.
Anon (2002) *Woodlands for Wales: the National Assembly for Wales Strategy for Trees and Woodlands Welsh Woodland Initiative*. Forestry Commission, Aberystwyth, UK.
Brunet, J. and von Oheimb, G. (1998) Migration of vascular plants to secondary woodlands in southern Sweden. *Journal of Ecology* 86, 429–438.
Cosgrove, P. (2002) *The Cairngorms Local Biodiversity Action Plan*. Cairngorms Partnership, Grantown on Spey, UK.
Currie, F. and Elliott, G. (1997) *Forests and Birds: a Guide to Managing Forests for Rare Birds*. RSPB, Sandy, UK.
Ferris, R., Peace, A.J., Humphrey, J.W. and Broome, A.C. (2000) Relationships between vegetation, site type and stand structure in coniferous plantations in Britain. *Forest Ecology and Management* 136, 55–83.
Forestry Commission (2002) *Forestry Statistics 2002*. Economic and Statistics Unit, Forestry Commission, Edinburgh.
Goudriaan, J. (1988) The bare bones of leaf angle distribution in radiation models for canopy photosynthesis and energy exchange. *Agricultural and Forest Meteorology* 43, 155–169.
Grashof-Bokdam, C.J. and Geertsema, W. (1998) The effect of isolation and history on colonization patterns of plant species in secondary woodland. *Journal of Biogeography* 25, 837–846.
Hamilton, G.J. (1998) *Forest Mensuration*. Forestry Commission Booklet No. 39. HMSO, London.

Hartley, M.J. (2002) Rationale and methods for conserving biodiversity in plantation forests. *Forest Ecology and Management* 155, 81–95.

Hermy, M., Honnay, O., Firbank, L., Grashof-Bokdam, C.J. and Lawesson, J.E. (1999) An ecological comparison between ancient and other forest plant species of Europe and the implications for forest conservation. *Biological Conservation* 91, 9–22.

Hersperger, A.M. and Forman, R.T.T. (2003) Adjacency arrangement affects plant diversity and composition in woodland patches. *Oikos* 101, 279–290.

Hill, M.O., Mountford, J.O., Roy, D.B. and Bunce, R.G.H. (1999) *Ellenberg's Indicator Values for British Plants*. ECOFACT, Vol. 2: Technical Annex. Centre for Ecology and Hydrology, Natural Environment Research Council, Cambridge, UK.

Hilmo, O. and Såstad, S.M. (2001) Colonization of old forest lichens in a young and old boreal *Picea abies* forest: an experimental approach. *Biological Conservation* 102, 251–259.

Honnay, O., Bossuyt, B., Verheyen, K., Butaye, J., Jacquemyn, H. and Hermy, M. (2002) Ecological perspectives for the restoration of plant communities in European temperate forests. *Biodiversity and Conservation* 11, 213–242.

Humphrey, J.W., Hawes, C., Peace, A.J., Ferris-Kaan, R. and Jukes, M.R. (1999) Relationships between insect diversity and habitat characteristics in plantation forests. *Forest Ecology and Management* 113, 11–21.

Humphrey, J.W., Newton, A.C., Peace, A.J. and Holden, E. (2000) The importance of conifer plantations in northern Britain as a habitat for native fungi. *Biological Conservation* 96, 241–252.

Humphrey, J.W., Ferris, R., Jukes, M.R. and Peace, A.J. (2001) Biodiversity in planted forests. In: *Forest Research Annual Report 2000–2001*. Forestry Commission, Edinburgh. pp 25–33.

Humphrey, J.W., Davey, S., Peace, A.J., Ferris, R. and Harding, K. (2002a) Lichen and bryophyte communities of planted and semi-natural forests in Britain: the influence of site type, stand structure and deadwood. *Biological Conservation* 107, 165–180.

Humphrey, J.W., Stevenson, A. and Swailes, J. (2002b) *Life in The Deadwood: a Guide to the Management of Deadwood in Forestry Commission Forests*. Forest Enterprise Living Forests Series. Forest Enterprise, Edinburgh.

Humphrey, J.W., Newton, A.C., Latham, J., Gray, H., Kirby, K.J., Quine, C.P. and Poulsom, E. (eds) (2003a) *The Restoration of Wooded Landscapes*. Proceedings of the conference 'Restoration of Wooded Landscapes', Heriot-Watt University, 12–14 September 2000. Forestry Commission, Edinburgh.

Humphrey, J.W., Ray, D., Watts, K., Brown, C., Poulsom, E.L., Griffiths, M. and Broome, A.C. (2003b) Balancing upland and woodland strategic priorities. Unpublished Contract Report to Scottish Natural Heritage AB(AA101)020398. Scottish Natural Heritage, Inverness, UK.

Humphrey, J.W., Ferris, R. and Quine, C.P. (eds) (2003c) *Biodiversity in Britain's Forests: Results From the Forestry Commission's Biodiversity Assessment Project*. Forestry Commission, Edinburgh.

Humphrey, J.W., Ferris, R. and Peace, A.J. (2003d) Relationships between site type, stand structure and plant communities. In: Humphrey, J.W., Ferris, R. and Quine, C.P. (eds) *Biodiversity in Britain's Forests: Results From the*

Forestry Commission's Biodiversity Assessment Project. Forestry Commission, Edinburgh.

Hunter, M.L., Jr (1990) *Wildlife, Forests and Forestry: Principles of Managing Forests for Biological Diversity.* Prentice-Hall, Englewood Cliffs, New Jersey.

Jukes, M.R., Peace, A.J. and Ferris, R. (2001) Carabid beetle communities associated with coniferous plantations in Britain: the influence of site, ground vegetation and stand structure. *Forest Ecology and Management* 148, 271–286.

Jukes, M.R., Ferris, R. and Peace, A.J. (2002) The influence of stand structure and composition on diversity of canopy Coleoptera in coniferous plantations in Britain. *Forest Ecology and Management* 163, 27–41.

Kirby, K.J. (1993) The effects of plantation management on wildlife in Great Britain: lessons from ancient woodland for the development of afforestation sites. In: Watkins, C. (ed.) *Ecological Effects of Afforestation.* CAB International, Wallingford, UK, pp. 15–30.

Kirby, K.J., Reid, C.M., Thomas, R.C. and Goldsmith, F.B. (1998) Preliminary estimates of fallen deadwood and standing dead trees in managed and unmanaged forests in Britain. *Journal of Applied Ecology* 35, 148–155.

McCullagh, P. and Nelder, J.A. (1989) *Generalised Linear Models,* 2nd edn. Chapman & Hall, London.

Moilanen, A. and Nieminen, M. (2002) Simple connectivity measures in spatial ecology. *Ecology* 83, 1131–1145.

Nature Conservancy Council (1986) *Nature Conservation and Afforestation in Britain.* Nature Conservancy Council, Peterborough, UK.

Peterken, G.F. (1993) *Woodland Conservation and Management,* 2nd edn. Chapman & Hall, London.

Peterken, G.F. (1996) *Natural Woodland Ecology and Conservation in Northern Temperate Regions.* Cambridge University Press, Cambridge, UK.

Petty, S.J., Garson, P.J. and MacIntosh, R. (ed.) (1995) Kielder: the ecology of a man-made spruce forest. Papers presented at a symposium held at the University of Newcastle-upon-Tyne, 20–21 September, 1994. *Forest Ecology and Management* 72(1–2).

Pyatt, D.G., Ray, D. and Fletcher, J. (2001) *An Ecological Site Classification for Forestry in Great Britain.* Forestry Commission Bulletin 124. Forestry Commission, Edinburgh.

Roberts, A.J., Russell, C., Walker, G.J. and Kirby, K.J. (1992) Regional variation in the origin, extent and composition of Scottish woodland. *Botanical Journal of Scotland* 46, 167–189.

Rodwell, J.S. (ed.) (1991) *British Plant Communities 1: Woodlands and Scrub.* Cambridge University Press, Cambridge, UK.

Spencer, J.W. (2002) *Ancient Woodland on the Forestry Commission Estate in England.* Forest Enterprise, Edinburgh.

Spencer, J.W. and Kirby, K.J. (1992) An inventory of ancient woodland for England and Wales. *Biological Conservation* 62, 77–93.

Verheyen, K. and Hermy, M. (2001) The relative importance of dispersal limitation of vascular plants in secondary forest succession in Muizen Forest, Belgium. *Journal of Ecology* 89, 829–840.

Vickers, A.D., Rotherham, I.D. and Rose, J.C. (2000) Vegetation succession and colonisation rates at the forest edge under different environmental conditions. *Aspects of Applied Biology* 58, 1–8.

Metapopulation Dynamics Following Habitat Loss and Recovery: Forest Herbs in Ancient and Recent Forests

12

M. Vellend

Department of Ecology and Evolutionary Biology, Cornell University, Ithaca, New York, USA

Many regions of Europe and eastern North America share broadly similar histories of land use over the past several centuries, with varying degrees of forest clearance followed by varying degrees of forest recovery. Here I use metapopulation models to explore the influence of the extent and timing of forest clearance and recovery on patch occupancy of slow-colonizing forest herbs in ancient and recent forests. If forest clearance exceeds the threshold for metapopulation persistence, extinction may require centuries to occur even when clearance is most severe. This suggests a heavy extinction debt for forest herbs in fragmented landscapes. The time between forest clearance and recovery (t_{c-r}) can have a strong influence on metapopulation dynamics. If t_{c-r} is short, patch occupancy in ancient forests (P_{AO}) will start out high, and exert a strong 'colonization pressure' on recent forests, increasing patch occupancy in recent forests (P_{RO}) above its equilibrium value before it ultimately declines towards equilibrium. If t_{c-r} is long, P_{AO} starts out low, and both P_{AO} and P_{RO} rise slowly towards equilibrium. The time required for P_{RO} to reach equilibrium is generally in the order of centuries, and increases with t_{c-r}. The results indicate that there may be long periods of time (decades to centuries) during which patch occupancy increases or decreases even in the absence of directional environmental change.

12.1 Introduction

In much of eastern North America, there have been two fairly distinct phases in the history of land use since European settlement. First was the widespread clearance of forest, largely during the 1800s, to create agricultural land. Forest cover was reduced during this phase from almost 100%, usually to less than 20%, depending on the particular region (Whitney, 1994). Not long after forest cover reached a minimum, sometime between the mid-19th and early 20th centuries, the second phase of land-use history was under way, during which time widespread abandonment of agricultural land allowed forest cover to increase; in some regions forest cover is now > 80% (Foster *et al.*, 1998).

In most of Europe, the history of intensive land use extends back in time for millennia rather than centuries, with the minimum forest cover < 5% in many regions (Kirby and Watkins, 1998). None the less, for many European regions it is possible to distinguish a relatively recent phase during which forest cover has increased (if only slightly) due to forest regrowth on former farmland (Honnay *et al.*, 2002a). Thus, many regions of Europe and eastern North America share broadly similar histories of forest-cover change over the past several centuries, with a phase of forest clearance followed by a phase of forest recovery. Forests that predate the earliest land-use maps in a given region are called 'ancient', and forests known to be growing on former agricultural land are called 'recent' (Rackham, 1980).

The influence of habitat loss and fragmentation on populations and communities has been the focus of a vast body of research in ecology and conservation biology. However, both theoretical and empirical studies have focused largely on the consequences of fragmentation for population dynamics in remnant habitat patches (Hanski, 1999; Haila, 2002), with little consideration of what happens when habitat recovers (but see Tilman *et al.*, 1997; Huxel and Hastings, 1999). This was recognized by Nee and May (1992), who stated that relative to habitat patch *removal*, 'the consequences of patch *addition* may be of more relevance to European ecology, if the farmland that is being removed from agriculture over the next decades is not simply paved over' (their italics). If remnant habitat patches provide the sources of colonists for recovered patches, it follows that the extent of habitat loss should influence re-establishment of populations and communities via the reduction of potential sources of colonists. In addition, for species with low rates of local colonization and extinction, such as many forest plants, there may be extended time lags between changes in landscape structure and changes in species distributions (Tilman *et al.*, 1994), such that we might expect not only the *extent* of habitat loss to be important, but also the relative *timing* of loss and recovery. In this chapter, I use metapopulation models to explore these issues for forest herbs in landscapes characterized by a history of forest clearance (i.e. habitat loss) and subsequent recovery.

During the past 20 years, an impressive body of knowledge has been accumulated concerning differences in plant community composition between ancient and recent forests in Europe (reviewed by Honnay *et al.*, 2002a) and eastern North America (Matlack, 1994; Singleton *et al.*, 2001; Bellemare *et al.*, 2002; Gerhardt and Foster, 2002). For forest herbs (i.e. herbaceous plant species whose distributions are restricted to forests), species diversity is generally lower in recent than in ancient stands. The relative paucity of forest herbs in recent stands can be explained by a combination of dispersal limitation and habitat limitation, although the bulk of the evidence appears to support dispersal limitation as the dominant process (Honnay *et al.*, 2002a). This points to an important aspect of forest herb ecology that is relevant to metapopulation modelling: rates of colonization of empty but suitable habitat patches are relatively low. In addition, because most forest herbs are long-lived perennials, rates of local extinction are also likely to be low.

Metapopulation models make predictions at the landscape scale. Dispersal-limited colonization suggests that landscapes with the lowest percentage cover of ancient forest should show the most severe reduction in forest herb diversity in recent forests, due to limited sources of colonists. This prediction was supported by a comparison of two landscapes in Belgium, in which colonization of recent forest was slower in the landscape with less ancient forest (Honnay *et al.*, 2002b). A meta-analysis of data from the literature for ten regions of Europe and eastern North America was also consistent with this prediction (Vellend, 2003; Fig. 12.1). The proportion of the landscape occupied by ancient forest had a strong influence, and the proportion of recent forest a relatively weak influence, on relative species diversity of ancient forest herbs in recent stands (see Fig. 12.1). This pattern was mirrored by patch occupancy patterns derived from a metapopulation model parameterized for a relatively poor colonizing species (Vellend, 2003). Here I modify the metapopulation model used in Vellend (2003) to explore two questions:

1. How does the extent of forest clearance influence the time to extinction? In other words, how much time do we have after forest clearance before forest recovery becomes necessary to prevent regional extinction of species?
2. If ancient and recent forests provide habitat of equivalent suitability for forest herbs, how does the length of time between forest clearance and recovery influence the length of time it takes for patch occupancy in recent stands to reach equilibrium? In other words, independent of the *extent* of forest clearance and recovery, how does the *timing* of these events influence metapopulation dynamics?

12.2 Metapopulation Models

Levins (1969) introduced the metapopulation concept, and formulated a simple model for an infinite number of equally connected, identical habitat

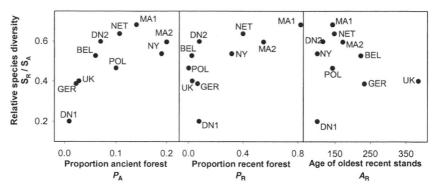

Fig. 12.1. Relationships among landscape variables in ten regions of Europe and eastern North America. Proportion ancient forest (P_A), proportion recent forest (P_R), and age of oldest recent stands (A_R) are plotted against the ratio of species diversity in recent vs. ancient stands (S_R/S_A) for ten landscapes of Europe and eastern North America: Flanders, Belgium (BEL); Røsnaes, Denmark (DN1); Himmerland, Denmark (DN2); Lower Saxony, Germany (GER); Petersham, Massachusetts (MA1); Shelburne and Conway towns, Massachusetts (MA2); Twente region, The Netherlands (NET); Tompkins County, New York (NY); Western Carpathian Foothills, Poland (POL); and Lincolnshire, United Kingdom (UK). (Reprinted with permission from Vellend, 2003, where details of the analysis and references can be found.)

patches in which the proportion of patches occupied (i.e. patch occupancy) is determined by a balance between local rates of colonization and extinction. A large number of far more complicated metapopulation models have been devised (Hanski, 1999), but surprisingly, the simple Levins model provides a very close approximation to 'spatially realistic' models in which the size and isolation of a large number of individual patches are treated explicitly (Hanski and Ovaskainen, 2002; Ovaskainen and Hanski, 2002). Here I use modifications of the Levins model to include two types of habitat patch, ancient forest and recent forest. The models I use are not novel contributions to the theoretical literature (see Hanski, 1999 for a wide range of modifications to the Levins model), but are novel as applied to forest herbs in ancient and recent forests.

Two assumptions of metapopulation models require comment. The first assumption is that a landscape can be treated as consisting of suitable habitat patches in a matrix of entirely unsuitable habitat. As such, the models used here apply only to species whose distributions are restricted more-or-less entirely to forests. Secondly, metapopulation models assume an important role for local extinction and colonization events. The objection may be raised that it is inappropriate to use such models for a group of species in which extinction and colonization events are extremely infrequent, and thus have little influence on short-term dynamics. However, all populations have a non-zero probability of extinction in any given year, and for understanding *long-term* dynamics, colonization and extinction processes are critical. Forest herbs are of particular interest in a conservation context precisely because of the slow rates at which these processes occur, and the resulting potential for extended time lags in response to landscape change.

12.2.1 Question 1: How does the extent of forest clearance influence time to extinction?

Following a modification of the Levins model that incorporates habitat loss (Hanski, 1999, p. 68 and references therein), the rate of change in patch occupancy in ancient forests (P_{AO}) can be modelled as follows:

$$\frac{dP_{AO}}{dt} = cP_A P_{AO}(1 - P_{AO}) - eP_{AO}$$

P_A is the proportion of the landscape that is forest after clearance ('A' denotes that this is ancient forest). The probability that the population in any given forest patch will go extinct per year is e, and the probability that any given empty patch will be colonized per year and per unit of occupied habitat is c. The definition of all symbols used in this chapter can be found in Table 12.1. Prior to forest clearance, the equilibrium patch occupancy is $P_{eq} = 1 - e/c$, whereas after clearance (and in the absence of any recovery) it is $P_{eq(clear)} = 1 - e/P_A c$. Therefore, if $P_A < e/c$ (i.e. the threshold for metapopulation persistence is crossed), the entire metapopulation will become extinct (Hanski, 1999). However, if e is low, the extinction process may be protracted, and a regional community may, for a long time, contain many species that are ultimately doomed to extinction. This phenomenon has been referred to as 'extinction debt' (Tilman et al., 1994), and may be particularly pertinent for forest herbs due to their potential for long-term local persistence. Using values of c and e estimated approximately for a relatively slow-colonizing forest herb species (see below), I used this model to ask how the extent of forest clearance ($1 - P_A$) influences time to extinction. Since in this model extinction is never actually reached, but only approached asymptotically, I define extinction as having occurred when $P_{AO} < 0.01$. For simplicity, I also assume that patch occupancy is at equilibrium prior to forest clearance, and that forest clearance happens entirely at one point in time. All modelling analyses were coded and run using MATLAB version 6.0 (The MathWorks Inc., Natick, Massachusetts, USA).

12.2.2 Question 2: How does the timing of forest clearance and recovery influence metapopulation dynamics?

In the above-described model, patch occupancy will decline following forest clearance, either to a new equilibrium (if $P_A > e/c$) or to extinction (if $P_A < e/c$). If forest regenerates, and habitat that was once destroyed becomes suitable

Table 12.1.   Definitions of symbols.

P_A	The proportion of a landscape that is ancient forest
P_R	The proportion of a landscape that is recent forest
P_{AO}	The proportion of ancient forest patches that are occupied
P_{RO}	The proportion of recent forest patches that are occupied
t_{c-r}	The time in years between forest clearance and recovery
c	The colonization rate per occupied patch
e	The extinction rate
P_{eq}	The equilibrium patch occupancy prior to forest clearance (since all forests at this time are ancient, this is the equilibrium value of P_{AO})
$P_{eq(clear)}$	The equilibrium patch occupancy following forest clearance (again, all forests at this time are ancient, so this is the equilibrium value of P_{AO})
$P_{eq(rec)}$	The equilibrium patch occupancy following forest recovery (since ancient and recent forest patches are identical, this is the equilibrium value of both P_{AO} and P_{RO}, although P_{AO} and P_{RO} can take different values before equilibrium is reached)
S_R/S_A	The ratio of species diversity of ancient forest herbs in recent vs. ancient stands

again, the overall equilibrium patch occupancy will rise. If we denote the proportion of the landscape that is recent forest as P_R, and assume that ancient and recent forests are equally suitable, the equilibrium patch occupancy following forest recovery is $P_{eq(rec)} = 1 - e/(P_A + P_R)c$. The proportion of recent patches that is occupied (P_{RO}) starts at zero following forest recovery, and the initial value of P_{AO} at the time of recovery (which is assumed to happen at one point in time) will depend on how much time has elapsed since forest clearance; both P_{RO} and P_{AO} will ultimately converge on $P_{eq(rec)}$. Thus, keeping track of patch occupancy in both recent (P_{RO}) and ancient (P_{AO}) patches, we can expect the dynamics to depend on the time between forest clearance and recovery. This may be relevant to understanding differences in metapopulation dynamics between European and North American landscapes, for which the differences in the times between forest clearance and recovery are large (long for Europe, short for North America).

To follow metapopulation dynamics in both recent and ancient forest patches requires solving the following two equations simultaneously.

$$\frac{dP_{AO}}{dt} = (cP_AP_{AO} + cP_RP_{RO})(1 - P_{AO}) - eP_{AO}$$

$$\frac{dP_{RO}}{dt} = (cP_AP_{AO} + cP_RP_{RO})(1 - P_{RO}) - eP_{RO}$$

I used this model to ask how the time between forest clearance and recovery influences the time it takes for patch occupancy in recent forest (P_{RO}) to reach equilibrium following recovery. I defined equilibrium as having been reached when P_{RO} first comes within 5% of the value of $P_{eq(rec)}$ (i.e. when $|P_{RO} - P_{eq(rec)}| < 0.05$).

12.2.3 Model parameterization

I used values of e based on literature reports. Repeated surveys of vascular plants on islands or forest patches in Europe have resulted in estimates of e ranging from 0.002 to 0.01 (Nilsson and Nilsson, 1985; Gibson, 1986; Jerling, 1998; Roden, 1998). These values represent averages across species and sites, and estimate the probability of any given species becoming extinct in any given patch or island in any given year. Because forest herbs are generally long-lived perennials, e is unlikely to be higher than 0.01.

In Vellend (2003), I used a model similar to those described above, but with ancient forests treated as a mainland where $e = 0$, to estimate the value of c needed to explain the observed value of $P_{RO} = 0.4$ about 80 years after forest recovery for *Aster divaricatus*, a slowly colonizing forest herb in central New York (data from Singleton *et al.*, 2001). With $e = 0.005$, the estimated value was $c = 0.039$.

I have explored model output for a range of parameter values, but for the sake of brevity I report results here using only $c = 0.039$ and $e = 0.005$ or 0.01. For Question 2, I report results for $P_R = P_A = 0.15$. In general, changing parameter values alter dynamics quantitatively but not qualitatively. Forest herb species vary widely in their colonization and persistence ability, and are thus likely to vary widely in their values of c and e. I consider the parameter values used here representative of a long-lived, relatively slowly colonizing species.

12.3 Results and Discussion

12.3.1 Fragmentation and metapopulation (regional) extinction

If the rate of local extinction is relatively low, as expected for most perennial forest herbs, the time to metapopulation extinction can be very long. With $e = 0.01$, the time to metapopulation extinction is > 400 years, even when forest clearance is severe (i.e. when $P_A < 0.01$), and up to several thousand years when the extent of clearance is only slightly below the metapopulation persistence threshold, which occurs at $P_A = e/c = 0.256$ (Fig. 12.2). The time to metapopulation extinction is longer with $e = 0.005$, although the qualitative relationship with P_A is similar to that for $e = 0.01$, with a sharp increase in time to extinction close to the persistence threshold (Fig. 12.2). This is the same result as that reported by Hanski and Ovaskainen (2002), but with parameters chosen here to represent a long-lived and slowly colonizing forest herb. Essentially there are two thresholds that pertain to metapopulation extinction. First,

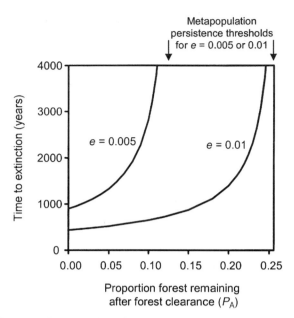

Fig. 12.2. Theoretical relationships between the proportion of forest remaining after fragmentation (P_A) and the time to metapopulation extinction when $c = 0.039$.

when $P_A < e/c$, extinction is certain but may take thousands of years to occur. When P_A is only slightly less than e/c, time to extinction shows a steep decline with P_A, and at $P_A \approx 0.8e/c$ (~0.1 for $e = 0.005$ and ~0.2 for $e = 0.01$) a second threshold is crossed below which further forest clearance has little influence on time to extinction.

A clear message from these results is that the extinction debt is likely to be substantial for forest herbs in many landscapes of Europe and eastern North America. Slow but steady decline in patch occupancy for many species may occur because of the slow crawl to extinction (or the slow decline to a positive patch occupancy equilibrium), even if the conditions for growth and reproduction within patches do not change. Because colonization and extinction rates vary among species, persistence thresholds for different species are likely to be crossed continually as forest clearance proceeds. However, even with only a small fraction of the landscape remaining as forest (e.g. $P_A < 0.01$), it may take centuries before many extinction events occur at the landscape scale. None the less, these extinctions are inevitable, at least theoretically, and the only ways they can be averted are if colonization rates are enhanced via seed or adult plant introductions, or if forests recover such that the proportion of forest in the landscape exceeds the persistence threshold. Because e is likely to be < 0.01 for many forest herbs, the conclusion that regional extinction will usually take centuries to occur, even with no changes to local habitat conditions, is probably quite robust.

12.3.2 Metapopulation dynamics and the time lag between forest clearance and recovery

Following forest clearance, patch occupancy declines from its initial equilibrium ($P_{eq} = 1 - e/c$) towards a new, lower equilibrium ($P_{eq(clear)} = 1 - e/P_Ac$). When forest recovery occurs, there is a new patch occupancy equilibrium ($P_{eq(rec)} = 1 - e/(P_A + P_R)c$), the value of which will be intermediate between the pre- and post-clearance values. In the unlikely case that the entire landscape becomes forested after recovery, it is possible that $P_{eq(rec)} = P_{eq}$. If recovery occurs shortly after forest clearance, it is likely that patch occupancy (P_{AO}) will initially be above $P_{eq(rec)}$, and so will continue to decline after recovery towards $P_{eq(rec)}$ (Fig. 12.3A, C). Patch occupancy in recent forest patches (P_{RO}) starts at zero, and if the initial P_{AO} is sufficiently high, P_{RO} may in fact first increase above $P_{eq(rec)}$ before converging on P_{AO} (Fig. 12.3C). P_{RO} and P_{AO} then decline together toward $P_{eq(rec)}$.

If there is a long time period between forest clearance and recovery, P_{AO} at the time of recovery is likely to have declined below $P_{eq(rec)}$ (Fig. 12.3B, D).

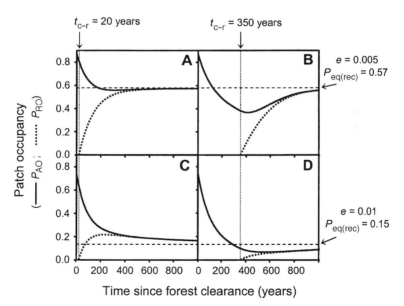

Fig. 12.3. Theoretical patch occupancy dynamics in ancient (solid curves) and recent (dotted curves) forests following forest clearance and recovery, with $e = 0.005$ (A, B) or $e = 0.01$ (C, D), and the time between clearance and recovery, $t_{c-r} = 20$ years (A, C) or $t_{c-r} = 350$ years (B, D). Forest clearance (occurring at time zero on the x-axis) and recovery were assumed to be instantaneous changes, with $P_A = P_R = 0.15$. Horizontal dashed lines indicate the equilibrium patch occupancy after recovery, $P_{eq(rec)}$, for the two extinction rates, and vertical dashed lines indicate t_{c-r}.

Interestingly, there may still be a period of 50+ years following forest recovery, during which P_{AO} continues to decline despite being below $P_{eq(rec)}$ (Fig. 12.3B, D). This is because the increase in forest cover due to recovery has little influence on overall patch occupancy until enough recent patches become occupied to act as significant sources of colonization. Once P_{RO} and P_{AO} converge, both increase towards $P_{eq(rec)}$.

The time it takes for P_{RO} to reach $P_{eq(rec)}$ increases with the time allowed to pass between forest clearance and recovery (t_{c-r}), though the shape and slope of the relationship depend strongly on the magnitudes of colonization and extinction rates (Fig. 12.4). The slope is much steeper when the extinction rate is relatively high ($e = 0.01$) than when the extinction rate is relatively low ($e = 0.005$; Fig. 12.4). The influence of t_{c-r} on the time to equilibrium for P_{RO} is mediated by the initial value of P_{AO} at the time of forest recovery. When recovery follows quickly after forest clearance, P_{AO} starts out high and thus exerts considerable 'colonization pressure' on the initially empty recent patches, thereby increasing the rate of approach of P_{RO} towards $P_{eq(rec)}$.

These results have two important implications. First, the timing of forest clearance and recovery can have dramatic effects on metapopulation dynamics, independently of the effects of the extent of clearance and recovery. Recovery of forests sooner rather than later can accelerate the increase in recent forest patch occupancy, in addition to preventing otherwise inevitable extinctions. Secondly, there is tremendous potential for extended time lags in the dynamics of forest herb metapopulations in changing landscapes.

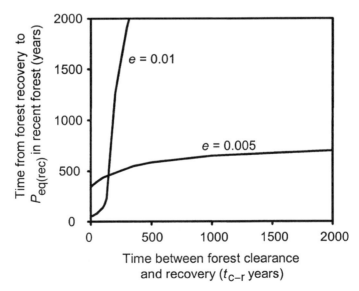

Fig. 12.4.   Theoretical influence of the time between forest clearance and recovery (t_{c-r}) on the time from recovery to patch occupancy equilibrium in recent forests.

Centuries, or even millennia, may be required for the regional distributions of forest herb species to reach equilibrium with landscape properties. When directional trends in landscape-level species abundance patterns are observed, it may be tempting to assume that there must be underlying directional change in local environmental conditions, human-caused or otherwise. The results here warn against such interpretations. Due to slow rates of local colonization and extinction, patch occupancy can change directionally for decades to centuries in the absence of continued landscape change. The slow rate of colonization of recent forests is well appreciated (Honnay *et al.*, 2002a,b), but other kinds of time lags are possible as well. For example, there may be long periods of time during which patch occupancy in both ancient and recent forests either increases (Fig. 12.3B) or decreases (Fig. 12.3C), depending only on the timing of forest clearance and recovery.

12.3.3 Models and data

Theoretical models are simplified representations of nature. In the models used here, many aspects of real plants growing in real landscapes have been ignored. In particular, there was no explicit treatment of spatial structure, habitat heterogeneity, habitat quality, the rate of change in landscape structure, or deterministic local extinction. These factors and others may all be important in determining the regional distributions and dynamics of species. However, as mentioned in the introduction, qualitative predictions from simple metapopulation models are unlikely to change with the addition of complexities such as variable patch characteristics.

It is when confronted with empirical data that simple models may run into trouble. Landscapes that vary in the extent and timing of forest clearance and recovery (e.g. North America versus Europe) are also likely to vary in a suite of other characteristics. For example, the duration of arable land use (i.e. the time between forest clearance and recovery) can have a direct influence on local soil properties (Verheyen *et al.*, 1999), and therefore an indirect influence on vegetation. This may obscure any effect of the timing of clearance and recovery *per se* on patch occupancy patterns. Despite these limitations, models are indispensable in the present context because they represent one of the few tools available for exploring the consequences of large-scale landscape changes on long-term population dynamics, and for placing empirical observations in a coherent framework. In the remainder of this chapter, I briefly discuss one set of empirical observations in the light of theoretical predictions regarding extinction debt, and then some promising lines of enquiry for the future.

Hanski and Ovaskainen (2002) have recently pointed out that a transient excess of species with low patch occupancies is a signature of extinction debt. Testing for such a signature requires data either from several time periods before and after forest clearance within a given landscape, or from multiple landscapes with different degrees of clearance. The expectation is that the excess of species with low patch occupancy would be greater where P_A is smaller. Since most studies report data for a single landscape, and the sampling regimes across studies are rarely identical, comparable data on multiple landscapes are few. However, Graae (2000) presented patch occupancy data for the same list of species in ancient forests in two landscapes in Denmark. In Himmerland and Hornsherred, the percentage of the landscape forested as of 1888 was 7% and 12%, respectively (Graae, 2000). Thus, we would predict a relative excess of rare species in Himmerland compared to Hornsherred, and this is exactly what was observed (Fig. 12.5). While recognizing that the extent of forest clearance is not the only variable that differs between the two land-scapes (Graae, 2000), these data suggest a greater extinction debt in the landscape with less ancient forest. If we define rare species as $0 < P_{AO} < 0.1$, there were considerably more rare species in Himmerland (33) than in Hornsherred (22), despite very little difference in the number of forest plant species present in the two landscapes (91 in Himmerland, 88 in Hornsherred).

   The extent and timing of forest clearance and recovery has varied greatly among different regions in the north-temperate zone (Whitney, 1994; Kirby and Watkins, 1998). For example, there has generally been more extensive clearance, longer periods of agricultural land use, and less recovery in Europe

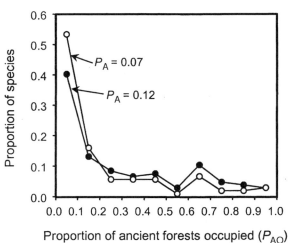

Fig. 12.5. The distribution of patch occupancies among species in two landscapes in Denmark [Himmerland (open circles), $P_A = 0.07$; and Hornsherred (closed circles), $P_A = 0.12$; data from Graae, 2000]. The proportion of species is presented in each of ten patch occupancy categories (0–0.1, > 0.1–0.2, etc.). There were significantly more rare species and fewer common species in Himmerland than in Hornsherred (Kolmogorov–Smirnov test, $P < 0.01$), regardless of whether or not absent species were included in the analysis (of 112 species searched for in both regions, 21 were absent in Himmerland and 24 in Hornsherred).

than in North America. Differences in soil characteristics between ancient and recent forest are also probably greater in Europe. As such, it is easy to predict that recolonization of forest plant species in recent forests should be far slower in Europe than in North America (Vellend, 2003), but much more difficult to pinpoint which particular aspects of landscape history drive such a pattern. When more comparable data become available for regional comparisons, the construction of more complex models may aid in distinguishing among historical hypotheses.

Two approaches to testing landscape-scale predictions in the future may prove particularly promising. There is now a wealth of data from many regions of Europe and North America derived from censuses of plant species in forest patches varying in age, area, isolation and environmental characteristics. If raw data are archived in an accessible format, with explicit information on geographic locations and sampling methods, researchers in the coming decades will be able to estimate extinction and colonization rates directly. Repeated surveys are available for a few individual sites (e.g. Gibson, 1986; Harmer *et al.*, 2001), but the potential exists for such valuable information to be collected in a range of different landscapes. Secondly, there is now the potential for direct comparisons among different landscapes. This requires careful selection of comparable landscapes, and statistically controlling for variation in multiple landscape characteristics, but these hurdles are surmountable, and this approach has already yielded some important insights concerning the long-term influence of habitat loss and fragmentation (Honnay *et al.*, 2002b; Vellend, 2003). Synthesis of results across regions and continents holds great potential for improving our understanding of how forest vegetation responds to continual landscape change.

Acknowledgements

I thank Peter Marks, Sana Gardescu, Kathryn Flinn, Jesse Bellemare, Jill Anderson, Stephen Ellner, and Parviez Hosseini for helpful comments, criticism, discussion or assistance with this project. Funding was provided by a training grant in Human Accelerated Environmental Change from the National Science Foundation, and a STAR fellowship from the US Environmental Protection Agency.

References

Bellemare, J., Motzkin, G. and Foster, D.R. (2002) Legacies of the agricultural past in the forested present: historical land-use effects on rich mesic forests. *Journal of Biogeography* 29, 1401–1420.

Foster, D.R., Motzkin, G. and Slater, B. (1998) Land-use history as long-term broad-scale disturbance: regional forest dynamics in central New England. *Ecosystems* 1, 96–119.

Gerhardt, F. and Foster, D.R. (2002) Physiographical and historical effects on forest vegetation in central New England, USA. *Journal of Biogeography* 29, 1421–1437.

Gibson, C.W.D. (1986) Management history in relation to changes in the flora of different habitats on an Oxfordshire estate, England. *Biological Conservation* 38, 217–232.

Graae, B.J. (2000) The effect of landscape fragmentation and forest continuity on forest floor species in two regions of Denmark. *Journal of Vegetation Science* 11, 881–892.

Haila, Y. (2002) A conceptual genealogy of fragmentation research: from island biogeography to landscape ecology. *Ecological Applications* 12, 321–344.

Hanksi, I. (1999) *Metapopulation Ecology*. Oxford University Press, Oxford, UK.

Hanski, I. and Ovaskainen, O. (2002) Extinction debt at extinction threshold. *Conservation Biology* 16, 666–673.

Harmer, R., Peterken, G., Kerr, G. and Poulton, P. (2001) Vegetation changes during 100 years of development of two secondary woodlands on abandoned arable land. *Biological Conservation* 101, 291–304.

Honnay, O., Bossuyt, B., Verheyen, K., Butaye, J., Jacquemyn, H. and Hermy, M. (2002a) Ecological perspectives for the restoration of plant communities in European temperate forests. *Biodiversity and Conservation* 11, 213–242.

Honnay, O., Verheyen, K., Butaye, J., Jacquemyn, H., Bossuyt, B. and Hermy, M. (2002b) Possible effects of habitat fragmentation and climate change on the range of forest plant species. *Ecology Letters* 5, 525–530.

Huxel, G.R. and Hastings, A. (1999) Habitat loss, fragmentation, and restoration. *Restoration Ecology* 7, 309–315.

Jerling, L. (1998) Linnaeus' Flora Kofsöensis revisited – floristic changes during 260 years in a small island of the lake Mälaren. *Nordic Journal of Botany* 18, 667–675.

Kirby, K.J. and Watkins, C. (1998) *The Ecological History of European Forests.* CAB International, Wallingford, UK.

Levins, R. (1969) Some demographic consequences of environmental heterogeneity for biological control. *Bulletin of the Entomological Society of America* 15, 237–240.

Matlack, G.R. (1994) Plant species migration in a mixed-history forest landscape in eastern North America. *Ecology* 75, 1491–1502.

Nee, S. and May, R.M. (1992) Dynamics of metapopulations: habitat destruction and competitive coexistence. *Journal of Animal Ecology* 61, 37–40.

Nilsson, I.N. and Nilsson, S.G. (1985) Experimental estimates of census efficiency and pseudoturnover on islands: error trend and between observer variation when recording vascular plants. *Journal of Ecology* 73, 65–70.

Ovaskainen, O. and Hanski, I. (2002) Transient dynamics in metapopulation response to perturbation. *Theoretical Population Biology* 61, 285–295.

Rackham, O. (1980) *Ancient Woodland: its History, Vegetation and Uses in England.* Edward Arnold, London.

Roden, C.M. (1998) Persistence, extinction and different species pools within the flora of lake islands in western Ireland. *Journal of Biogeography* 25, 301–310.

Singleton, R., Gardescu, S., Marks, P.L. and Geber, M.A. (2001) Forest herb colonization of postagricultural forests in central New York State. *Journal of Ecology* 89, 325–338.

Tilman, D., May, R.M., Lehman, C.L. and Nowak, M.A. (1994) Habitat destruction and the extinction debt. *Nature* 371, 65–66.

Tilman, D., Lehman, C.L. and Kareiva, P. (1997) Population dynamics in spatial habitats. In: Tilman, D. and Kareiva, P. (eds) *Spatial Ecology*. Princeton University Press, Princeton, New Jersey, pp. 3–20.

Vellend, M. (2003) Habitat loss inhibits recovery of plant diversity as forests regrow. *Ecology* 84, 1158–1164.

Verheyen, K., Bossuyt, B., Hermy, M. and Tack, G. (1999) The land use history (1278–1990) of a mixed hardwood forest in western Belgium and its relationship with chemical soil characteristics. *Journal of Biogeography* 26, 1115–1128.

Whitney, G.G. (1994) *From Coastal Wilderness to Fruited Plain: a History of Environmental Change in Temperate North America from 1500 to the Present.* Cambridge University Press, Cambridge, UK.

Short-term and Mid-term Response of Ground Beetle Communities (Coleoptera, Carabidae) to Disturbance by Regeneration Felling

13

E. Richard, F. Gosselin and J. Lhonoré*

Cemagref – Agricultural and Environmental Engineering Research, Ecosylv Team: 'Biodiversity and lowland forest management', Domaine des Barres, Nogent sur Vernisson, France

In a temperate, deciduous even-aged forest context we tested two main hypotheses related to the short-term and mid-term response of carabid communities to regeneration felling:

1. The 'perturbation hypothesis': species abundances change in opposite directions between species shortly after felling (maximum 20 years).
2. The 'succession hypothesis': species abundances peak, or bottom out, at different silvicultural stages depending on the species (up to 100 years).

Our results confirmed the two hypotheses. Moreover, the species that were significantly affected negatively in the short term by the regeneration felling (especially *Carabus nemoralis* and *C. violaceus*) increased in abundance along the first 100 years of the silvicultural cycle, suggesting that the regeneration felling did not have a detrimental effect on carabid communities on a 100-year time scale.

* Professor at the University of Maine, France, in delegation to Cemagref – Agricultural and Environmental Engineering Research.

©CAB International 2004. *Forest Biodiversity: Lessons from History for Conservation*
(eds O. Honnay, K. Verheyen, B. Bossuyt and M. Hermy)

13.1 Introduction

In Europe, the sustainable management of forests implies the reconciliation of timber production and biodiversity conservation (Third Ministerial Conference on the Protection of Forests in Europe, 1998). Therefore, the impact of silvicultural practices on biodiversity needs to be evaluated. In large productive forests, the main harvesting method is clear-cutting of mature, even-aged stands. Yet, the results of previous research in boreal coniferous forests on the short-term effects of this practice on carabid communities are quite alarming (Niemelä *et al.*, 1993; Beaudry *et al.*, 1997; Duchesne *et al.*, 1999; Koivula, 2002). Even if total carabid abundance and species richness increase shortly after felling, this is mainly due to the invasion by open-habitat species and the persistence of forest-generalist species, whereas specialist species of old-growth forests are reduced in abundance. Moreover, Niemelä *et al.* (1993) found no indication of the recovery of these negatively affected species 27 years after clear-cutting.

In other cases, and especially in some large productive state forests in France, mature stands of even-aged high forest are harvested through progressive regeneration felling (*sensu* Helms, 1998). In regeneration felling, scattered seed trees are first retained to fructify and to produce a new age class (seeding cut). As soon as the seedlings are established, the seed trees are then gradually removed through secondary fellings. Also, in the oak forest that is studied here, the whole upper tree layer is removed within 8–12 years after the first felling, when the seedlings reach *c.* 80 cm in height (ONF, 1997).

We formulated the following questions with respect to the short-term and mid-term effects of regeneration felling in a temperate oak forest:

1. Does the regeneration felling induce patterns of perturbation (*sensu* Rykiel, 1985; Blondel, 1995) in the short term (up to 20 years)? We tested whether the patterns found in a temperate, deciduous forest were similar to those observed in boreal coniferous forests (Niemelä *et al.*, 1993; Beaudry *et al.*, 1997; Duchesne *et al.*, 1999; Koivula, 2002) by synchronously comparing the abundance of each carabid species before and after the regeneration felling. Very specifically, we tested the following hypothesis: 'Species react oppositely to the felling: some species decrease in abundance, others increase and others remain steady' (i.e. H1, the perturbation hypothesis).
2. If there are changes in species abundance, do these changes last until the mid-term (up to 100 years), or does the regeneration felling induce patterns of succession (*sensu* Begon *et al.*, 1996) during the subsequent silvicultural cycle? We tested the following specific hypothesis: 'Species differ regarding the silvicultural stage where they peak (or bottom out)' by synchronously comparing carabid species abundances between the silvicultural stages of the first half of an oak even-aged cycle (i.e. H2, the succession hypothesis).

13.2 Material and Methods

13.2.1 Study area and sampling design

The forest under study (Montargis state forest, located 130 km south of Paris) is a large (4000 ha) ancient forest, the limits of which have remained nearly unchanged since the 12th century (Garnier, 1965). The majority of the stands are dominated by *Quercus petraea* and *Q. robur*, with an understorey layer composed mainly of *Carpinus betulus* and *Fagus sylvatica* in varying proportions. The main management goal is to produce quality timber for construction. Therefore, 70% of the area is managed as an oak even-aged high forest. In this system, the silvicultural cycle originates from progressive regeneration felling and is planned to last 200 years in order to allow the trees to reach 80 cm in diameter (ONF, 1996). The soil conditions are relatively homogeneous, with plateau soils on chalk substrate. There are small ranges in soil texture (sandy to silt–sandy), soil depth (40–70 cm), pH (slightly acid to acid) and stone content (little to average) (Chevalier, 2003).

In order to evaluate the response of the ground beetle communities to regeneration felling, we built a 68-plot sampling design aimed at allowing a synchronous comparison of different silvicultural stages (Table 13.1). To be able to test the hypotheses separately for the short-term and mid-term responses of ground beetle communities to regeneration felling, we divided the established plots into two different data sets (Table 13.1).

Table 13.1. Silvicultural characteristics of the stands in the sampling design. The first sampling design (bold) corresponds to the pre-felling and short-term post-felling stages. The second sampling design (not bold) corresponds to the subsequent silvicultural cycle.

	Pre-felling: old coppice with standards under conversion	Post-felling: even-aged forest — Short-term — (1) seeding cut	Short-term — (2) sapling stand	Mid-term — (3) pole stage	Mid-term — (4) high pole wood	Mid-term — (5) young high forest
Number of plots	**16**	**16**				
		4	12	12	12	12
Time since the last large felling (years)	**105**	**14**				
		1	18	37	50	90
Top height (m)	**26.1**	**5.2**				
		6.6	4.7	14.2	17.8	24.9
Mean basal area (m²/ha)	**27.8**	**9.6**				
		7.3	10.4	26.7	26.5	25.3

The 'perturbation' hypothesis (H1) was tested through the synchronous comparison of carabid communities before and after the regeneration felling (32 plots). The 16 plots in the mature pre-felling stage consisted of old coppice with standards under conversion to even-aged high forest. The 16 plots in the post-felling stage consisted of 4 stands just submitted to a seeding cut and 12 sapling stands originating from natural regeneration.

The 'succession' hypothesis (H2) was tested through the synchronous comparison of carabid communities between five different silvicultural stages (52 plots from 1 to 90 years after regeneration felling), corresponding to the first half of an oak even-aged high forest cycle in this area: (i) seeding cut (4 plots); (ii) sapling stands (12 plots); (iii) pole stage (12 plots); (iv) high polewood (12 plots); and (v) young high forest (12 plots).

13.2.2 Data collection: pitfall-trapping of ground beetles

Carabids were collected in pitfall traps in 1999, during three 1-week periods only (the second and fifth week of June and the fifth week of August). Niemelä *et al.* (1990) suggested that samples obtained by trapping periods of 20 days or more were similar enough to the whole-season samples to be used in several types of ecological studies, especially those that focus on individual species. In each plot, four traps arranged in a 14×14 m square were established. The traps (8.5 cm in diameter $\times$ 11 cm in depth) were half-filled with a solution of 50% ethylene glycol and sheltered by a transparent plastic roof in order to prevent flooding by rain. Ground beetles were identified according to Jeannel (1941), Lindroth (1974) and Hurka (1996). The nomenclature followed Hurka (1996). The larvae were not identified and were therefore analysed as a separate group.

13.2.3 Data analysis: multi-species ANOVA models

For each sampled plot, the catches from all traps and all periods were pooled. Because of trapping incidents, we did not keep the same number of traps per plot. Therefore, we worked with the standardized total abundance of species per plot across the three 1-week periods of trapping, defined as: y_{ijk} = mean abundance per active trap of species i in plot k (and stage j) $\times$ total number of traps placed in each plot.

Species with fewer than five occurrences in the three 1-week periods were summed according to their habitat preference, as defined by Desender (1986) and Coulon *et al.* (2000). Hence, they were analysed as separate groups of non-frequent species sharing similar habitat preferences.

To test the community and the species response within the same model, we used an original approach, finalized by Gosselin (unpublished). It consisted of rearranging the two-dimensional plot-vs.-species data set into a one-dimensional list-wise data set, in which the abundances observed for every species in every plot were put in the same column. The identity of the species and the environmental characteristics of the plots were presented in two separate columns. In this way, the observed abundances could be explained by both the identity of the species and the silvicultural variables, using a parametric model.

To test the short-term response of the whole community to felling (H1 with the first data set), we used a linear mixed-effect model (Pinheiro and Bates, 2000) which considered that the abundance of one species observed in one plot corresponding to a particular stage depended: (i) on the identity of the species; (ii) on the silvicultural stage of the plot; and (iii) on the particular response of this species to the stage:

$$y_{ijk} = a_{ijk}(\beta 1 + \beta 2\ \text{species}_i + \beta 3\ \text{stage}_j + \beta 4\ \text{species}_i \times \text{stage}_j) + \varepsilon_{ijk}$$

where y_{ijk} = vector of standardized abundances of species$_i$ in plot$_k$ and stage$_j$; a_{ijk} = reference abundance: predicted by an edaphic model established in other adult stands or estimated in these stands (Richard, unpublished).

A location term (part of the forest) was included in the error term (ε_{ijk}) as a random effect to eliminate a possible spatial bias in the sampling design. The non-normality of the residual distribution was not considered to be an obstacle, since White and Bennetts (1996) showed that ANOVA is a robust method for analysing overdispersed data. On the other hand, heteroscedasticity was modelled using variance functions depending on the identity of the species and on its level of abundance, and also on the number of actual traps considered for each plot. Thus, the existence of a perturbation pattern was first tested at the whole community level through the significance of the effects in the model: the significance of the main 'stage' effect means that there was a change in carabid abundance after regeneration felling, while the significance of the interaction term 'species × stage' means that the felling effect varied between species.

Secondly, to describe the species-specific response to the felling, we compared their predicted mean abundance before and after the felling, performing Scheffé *post-hoc* tests. Species were assigned to different perturbation profiles according to the significance of the pairwise comparison test:

- if the difference was not significant, the species was considered to be 'unaffected' by the felling;
- if the species had significantly higher abundance before the felling than after, the species was considered to be 'negatively affected' by the felling;
- if the species had significantly lower abundance before the felling than after, the species was considered to be 'favoured' by the felling.

Profiles were characterized through the number of species or groups that belonged to each profile at the 0.05 level of significance. If this number was different from zero, the profile was considered to exist. To assign a particular species or group to a profile with more certainty, the significance level was set at 0.01. Indeed, as the multiple comparisons were performed separately for each species or group, the significance level of 0.05 implies that one out of 20 species could be assigned to a wrong profile because of chance events. To avoid such a misinterpretation, we lowered the significance levels when examining the responses of individual species.

The same approach was then used to test the 'succession' hypothesis (H2) with the second data set. Species successional profiles over the first 100 years of the cycle were defined by the significance of the comparisons of their predicted mean abundance in the different silvicultural stages (Tukey tests):

- 'unaffected' species: none of the comparisons was significant;
- 'pioneer' species: significant maximum of the predicted mean abundance in the first stages (stage 1 or 2 had significantly higher abundance than at least two stages among 3, 4 or 5);
- 'non-late' species: significant minimum in the last stages (stage 4 or 5 had significantly lower abundance than at least one stage between 1 or 2);
- 'non-pioneer' species: significant minimum in the first stages (stage 1 or 2 had significantly lower abundance than at least one stage among 3, 4 or 5);
- 'intermediate' species: significant maximum in intermediate stages (stage 3 or 4 had significantly higher abundance than stage 1 or 2 and stage 5);
- 'late' species: significant maximum in the last stages (stage 4 or 5 had significantly higher abundance than at least two stages among 2, 3 or 4).

13.3 Results

A total of 12,445 individuals were collected (5973 for the first dataset and 10,451 for the second). *Carabus auratus* abundance comprised half of the total catch in the first dataset, and one-third of the second dataset. The other abundant species and species groups were *Abax parallelipipedus*, *C. problematicus*, *Pterostichus madidus*, and carabid larvae.

Forty-seven species were identified. After grouping the non-frequent species according to habitat preference, we selected 22 species, or groups of species, to test the 'perturbation' hypothesis (H1) and 27 species, or groups of species, to test the 'succession' hypothesis (H2).

13.3.1 Short-term response of the carabid community to the regeneration felling

Table 13.2 gives the results of the multi-species ANOVA that tested the 'perturbation' hypothesis. The significance of the main 'species' effect was trivial: it only confirmed that species differed from one another in their total abundance in the whole dataset. On the other hand, the high significance of the main 'felling' effect shows that the total abundance of carabids changed after regeneration felling. Furthermore, the interaction term 'species × stage' was also highly significant, i.e. the response varied between species.

The species-specific responses are reported in Table 13.4, which shows that each of the three perturbation profiles ('unaffected', 'negatively affected' or 'favoured') was represented by at least one species. Therefore, the 'perturbation' hypothesis was confirmed: 'species react oppositely to the felling'. Eight species did not react significantly to the regeneration felling and were therefore classified as 'unaffected' species: *Abax ovalis, A. parallelus, C. problematicus, Calathus rotundicollis, Pterostichus niger, P. oblongopunctatus,* a group of non-frequent forest species and a group of ubiquitous non-frequent species. Six species were classified as 'negatively affected' species, among which were *C. nemoralis* and *C. violaceus.* Eight species or groups appeared to be 'favoured' by the regeneration felling, particularly *C. auratus, Amara lunicollis, P. madidus, Poecilus* species, a group of non-frequent species of open habitats and a group of non-frequent species of dry habitats.

13.3.2 Mid-term response of the carabid community to regeneration felling

Table 13.3 reports the results of the multi-species ANOVA that tested the 'succession' hypothesis (H2). The main effect of the silvicultural stage is highly significant, which means that the total abundance of carabids changed during the first half of the cycle. The interaction term 'species × stage' is also highly

Table 13.2. Short-term effects of regeneration felling on the carabid community (multi-species ANOVA).

Effects	Numerator df	Denominator df	F value	P value
Species	21	629	11.870	< 0.0001
Felling stage	1	629	40.804	< 0.0001
Species × stage	21	629	6.150	< 0.0001

Table 13.3. Mid-term effects of regeneration felling on the carabid community (multi-species ANOVA).

Effects	Numerator df	Denominator df	F value	P value
Species	27	1218	17.570	< 0.0001
Silvicultural stage	4	1218	5.203	0.0004
Species × stage	104	1218	5.736	< 0.0001

Table 13.4. Species-specific response to regeneration felling, in the short term and the mid-term: *post-hoc* comparison tests of their predicted means per stage.

| Species/ groups | Short-term species response: | | Predicted difference: H1: | | Mid-term species response: | | | | | Predicted differences: | H2: |
| | Observed mean abundance | | | | Observed mean abundance | | | | | | Succession profile |
	Before felling	After felling	Scheffé tests	Perturbation profile	1	2	3	4	5	Tukey tests	
Abax ovalis (Duft.)	4.8	4.1	NS	Unaffected	6.0	3.4	16.0	7.3	4.7	3 > 1**.2***.4*.5**	Intermediate
Abax parallelepipedus (Pill. & Mitt)	39.0	16.3	*	(Negatively affected)	14.5	16.9	67.0	54.6	40.7	3*** > 1.2 & 4*** > 1.2 & 5* > 1.2	Non-pioneer
Abax parallelus (Duft.)	3.4	1.3	NS	Unaffected	2.8	0.8	3.5	3.0	8.6	5 > 2***	Non-pioneer
Amara lunicollis (Schiødte)	0.0	1.1	**	Favoured	2.8	0.6	0.0	0.0	0.0	1 > 2*.3**.4**.5** & 2 > 3**.4**.5**	Pioneer
Calathus luctuosus (Latreille)	0.3	2.6	*	(Favoured)	4.0	2.2	1.5	3.7	5.2	NS	Unaffected
Calathus rotundicollis (Dejean)	0.7	5.0	NS	Unaffected	18.3	0.6	0.2	3.8	6.1	5 > 2*	(Non-pioneer)
Carabus auratus (L.)	18.9	156.8	****	Favoured	71.0	185.5	54.0	41.1	4.4	2 > 1***.3***.4****. 5**** & 1.3***.4**.4** > 5	Pioneer
Carabus auronitens (F.)	1.3	0.1	*	(Negatively affected)	0.5	0.0	0.5	1.2	2.6	NS	Unaffected
Carabus nemoralis (O.Mül.)	10.2	1.0	****	Negatively affected	2.8	0.4	1.1	8.6	17.3	5 > 1**.2***.3*** & 4 > 2**.3*	Late
Carabus problematicus (Herbst)	17.0	10.3	NS	Unaffected	27.5	4.5	15.0	32.6	19.2	3**.4****.5** > 2	Non-pioneer
Carabus violaceus (F.)	12.9	2.1	***	Negatively affected	4.0	1.5	5.0	14.2	8.7	4 > 2***.3** & 5 > 2**	Late
Harpalus latus (L.)	0.0	1.4	*	(Favoured)	0.8	1.6	0.0	0.2	0.0	2 > 3*.5*	(Pioneer)
Leistus rufomarginatus (Duft.)	1.1	0.1	*	(Negatively affected)	0.3	0.1	0.1	1.1	1.4	5 > 2*.3*	(Late)

Species				Stage 1	Stage 2	Stage 3	Stage 4	Stage 5		Response
Molops piceus (Panz)	–	–	–	–	0.3	0.0	0.3	0.1	0.0	NS	Unaffected
Platyderus ruficollis (Marsh.)	–	–	–	–	0.0	0.0	0.4	0.1	0.0	NS	Unaffected
Pterostichus madidus (F.)	0.8	22.9	***	Favoured	10.8	26.9	12.8	10.8	3.4	2 > 1**. 3*. 4*. 5**** & 3*. 4* > 5	Non-late (pioneer)
Pterostichus niger (Schaller)	0.2	1.3	NS	Unaffected	4.8	0.1	2.7	2.7	0.3	NS	Unaffected
Pterostichus oblongopunctatus (F.)	1.1	0.2	NS	Unaffected	0.8	0.0	0.0	1.0	2.9	5 > 2**. 3**	Late
Synuchus vivalis (Ill.)	–	–	–	–	5.8	0.8	0.8	0.7	0.0	1 > 5*	(Non-late)
Carabid larvae	14.2	8.9	*	(Negatively affected)	18.0	5.9	13.4	13.1	18.2	3*. 4*. 5** > 2	Non-pioneer
Poecilus species	0.0	7.5	**	Favoured	12.5	5.9	0.2	0.0	0.0	2 > 3*. 4*. 5*	(Pioneer)
non-frequent forest or ubiquitous species s.l.	0.4	0.5	NS	Unaffected	–	–	–	–	–	–	–
non-frequent forest or ubiquitous species s.s.	–	–	–	–	0.0	0.0	0.4	0.6	0.2	NS	Unaffected
non-frequent species of dry habitats	0.0	1.0	**	Favoured	0.3	1.3	0.0	0.1	0.0	2 > 3*. 4*. 5**	Pioneer
non-frequent species of open habitats	0.0	0.8	**	Favoured	0.0	1.1	0.0	0.0	0.0	2 > 1****. 3****. 4****. 5****	Pioneer
non-frequent ubiquitous species s.l.	0.3	1.5	NS	Unaffected	–	–	–	–	–	–	–
non-frequent ubiquitous species s.s.	–	–	–	–	2.3	0.9	0.0	0.0	0.0	1 > 3*. 4*. 5* & 2 > 3*. 4**. 5**	Pioneer
non-frequent ubiquitous or forest species	–	–	–	–	1.3	0.0	0.1	0.0	0.1	NS	Unaffected
non-frequent ubiquitous or openland species	–	–	–	–	0.5	0.0	0.0	0.1	0.3	NS	Unaffected

The response profiles are indicated between brackets when the criteria are fulfilled at an 0.05 level of significance, and without brackets at an 0.01 level of significance. '3 > 1**. 2***. 4*. 5****' means that the predicted mean abundance in stage 3 (pole stage) is significantly higher than those in stage 1 (with $P < 0.01$), in stage 2 ($P < 0.001$), in stage 4 ($P < 0.05$) and in stage 5 ($P < 0.0001$).

significant: each species showed specific dynamics through the silvicultural stages.

Each of the six successional profiles: 'unaffected', 'non-late', 'pioneer', 'non-pioneer', 'intermediate' and 'late' was represented by at least one species (Table 13.4). Therefore, the H2 hypothesis was also confirmed. Eight species or groups did not respond significantly to the silvicultural stages and were therefore classified as 'unaffected' species: *Calathus luctuosus*, *Carabus auronitens*, *Molops piceus*, *Platyderus ruficollis*, *Pterostichus niger*, two groups of forest or ubiquitous non-frequent species, and a group of open-land or ubiquitous non-frequent species. Eight species or groups belonged to the 'pioneer' group, among which *C. auratus*, *A. lunicollis*, a group of non-frequent species of dry habitats, a group of non-frequent species of open habitats and a group of non-frequent ubiquitous species. Two species fell into the 'non-late' group, particularly *P. madidus*. Five species or groups were 'non-pioneer' species, particularly *A. parellelepipedus*, *A. parallelus*, *C. problematicus*, and the carabid larvae. *A. ovalis* was the only 'intermediate' species observed. The 'late' species group included four species, among which were *C. nemoralis*, *C. violaceus* and *P. oblongopunctatus*.

13.4 Discussion

Our results in the Montargis forest were in accordance with the 'perturbation' hypothesis. In the first stage, the entire carabid community was perturbed (*sensu* Rykiel, 1985; Blondel, 1995) by the regeneration felling, since carabid abundance changed on a 20-year time scale. Secondly, the response varied between species: they fell into three perturbation profiles – some species decreased in abundance, others increased and others remained steady. Therefore, we found that the perturbation showed the same patterns as those observed in some boreal coniferous forests after clear-cutting (Niemelä *et al.*, 1993; Beaudry *et al.*, 1997; Duchesne *et al.*, 1999; Koivula, 2002). However, other studies in boreal coniferous forests showed the existence of only two response profiles (Atlegrim *et al.*, 1997; Abildsnes and Tømmerås, 2000). Atlegrim (1997) found that most of the species were unaffected, with only one species significantly peaking in the clear cuts. This study, however, was based on only 179 individuals representing nine species. On the contrary, Abildsnes and Tømmerås (2000) did not find any species 'favoured' in the clear-cut fragmented areas as compared to the old-growth control, but only the seven most common species were investigated among the 18 that were captured. In addition, only the carabids from unlogged parts of fragmented areas versus control areas were compared. Moreover, as they made a diachronical study, the criterion used to define the species profile was different from ours; a synchronical analysis could have led to different results.

The two species strongly 'negatively affected' by the regeneration felling ($P < 0.01$) both belonged to the *Carabus* genus: *C. violaceus* and *C. nemoralis*. Both are also more abundant in a mature deciduous beech forest as compared to a 5-year-old spruce plantation in clear-cuts, according to Magura *et al.* (2003). Yet, these two *Carabus* species do not seem to have strict micro-environment requirements (Koch, 1989), since they are eury-thermic and eury-hygric species (Thiele, 1977). Furthermore, they did not completely disappear in the post-felling stage. Nevertheless, they both prefer dark conditions (Thiele, 1977). *C. violaceus* seems to be more forest-specific, since it occurs more frequently in 'ancient' forests (wooded continuously since the end of the 18th century) than in recent ones (Assmann, 1999). *C. nemoralis*, on the other hand, can reproduce either in woodlands or in open land (Assmann, 1999). Among the eight species or groups that increased in abundance after the regeneration felling, five were completely absent from the pre-felling stage [*A. lunicollis*, *Poecilus* species, a group of non-frequent open-habitat species, a group of non-frequent dry habitat species and *Harpalus latus* (hardly significant)]. Therefore, colonization occurred by some new, non-forest species (*sensu* Desender, 1986; Coulon *et al.*, 2000) within the 20 years after the regeneration felling.

Our results also agreed with the 'succession' hypothesis (H2). First, carabid communities showed successional patterns (*sensu* Begon *et al.*, 1996) during the first half of an even-aged high forest cycle (100 years), since the carabid abundance changed, depending on the silvicultural stage of the stand. Secondly, as described in H2, the species did not all react similarly: species abundance peaked or bottomed out at different silvicultural stages. Six different successional profiles were observed: during the first half of the silvicultural cycle, some species increased more or less rapidly ('non-pioneer' or 'late' species), others decreased ('non-late' or 'pioneer' species), while others increased then decreased ('intermediate') or remained rather steady ('unaffected'). These patterns resemble those of a 'relay' succession (derived from Egler, 1954) with alternating appearance and disappearance of groups of species (Oliver and Larson, 1996, p. 146), more than those of a 'nested' succession, in which one single stage would include all the species, and the other stages would be subsets of the most species-rich community (see Worthen, 1996). In our case, this interpretation was based only on the maximum or minimum abundance and not on the real appearance and disappearance of species, but we also observed that the minima sometimes corresponded to a *quasi-nul* catch of the species. In boreal coniferous forests (Niemelä *et al.*, 1993; Koivula *et al.*, 2002) and in temperate coniferous plantations (Baguette and Gerard, 1993; Butterfield, 1997; Magura *et al.*, 2003), others have also observed patterns of 'relay' succession based on abundance analyses. However, they only distinguish 'open-canopy' communities from 'closed-canopy' ones. This simple successional response of communities may be linked to the short length of the coniferous silvicultural cycles (maximum amplitude of the studied cycles: 40–70 years) which creates contrasting environmental conditions between young and mature stands.

To summarize our results, we can compare the short-term and mid-term responses of each species to felling. On the one hand, we checked that the species that were highly 'negatively affected' by the felling in the short term (*C. nemoralis* and *C. violaceus*) were found again in the subsequent silvicultural cycle with a maximum abundance in the latest sampled stages; this probably means that these species started to recover from regeneration felling about 80 years after disturbance. Even the species or groups that were less significantly 'negatively affected' (*A. parallelepipedus*, *C. auronitens*, *Leistus rufomarginatus* and carabid larvae) also increased in abundance, except for *C. auronitens* whose increase was not significant, possibly because of the edaphic correction which did not seem fully adapted to the species (Richard, unpublished). On the other hand, the species that were favoured by the regeneration felling either seemed to decrease rather slowly (*C. auratus*, *P. madidus*) or to decrease more rapidly (*A. lunicollis* or groups of non-frequent species).

13.5 Conclusions

To conclude, from a conservation point of view, our results are rather positive since the 'relay' succession allowed the coexistence of different species along the oak high-forest cycle, and since the negatively affected species seemed to start to recover, at least in the latest observed stages. Surprisingly, however, the more clearly 'negatively affected' species were rather forest-generalist species. We should therefore keep in mind that the species most sensitive to felling (unable to recover within 100 years) might have already disappeared from our forested landscapes because of their long history of disturbance.

References

Abildsnes, J. and Tømmerås, B.A. (2000) Impacts of experimental habitat fragmentation on ground beetles (*Coleoptera, Carabidae*) in a boreal spruce forest. *Annales Zoologici Fennici* 37, 201–212.

Assmann, T. (1999) The ground beetle fauna of ancient and recent woodlands in the lowlands of north-west Germany (*Coleoptera, Carabidae*). *Biodiversity and Conservation* 8, 1499–1517.

Atlegrim, O., Sjoberg, K. and Ball, J.P. (1997) Forestry effects on a boreal ground beetle community in spring: selective logging and clear-cutting compared. *Entomologica Fennica* 8, 19–26.

Baguette, M. and Gerard, S. (1993) Effects of spruce plantations on carabid beetles in southern Belgium. *Pedobiologia* 37, 129–140.

Beaudry, S., Duchesne, L.C. and Cote, B. (1997) Short-term effects of three forestry practices on carabid assemblages in a jack pine forest. *Canadian Journal of Forest Research* 27, 2065–2071.

Begon, M., Harper, J.L. and Townsend, C.R. (1996) *Ecology: Individuals, Populations and Communities*. Blackwell Scientific Editions, Oxford, UK.

Blondel, J. (1995) *Biogéographie. Approche écologique et évolutive*. Masson, Paris.

Butterfield, J. (1997) Carabid community succession during the forestry cycle in conifer plantations. *Ecography* 20, 614–625.

Chevalier, R. (2003) *Sylviculture du Chêne et Biodiversité Végétale Spécifique: Étude d'une Forêt en Conversion vers la Futaie Régulière: La Forêt Domaniale de Montargis (45)*. Cemagref, Nogent sur Vernisson, France.

Coulon, J., Marchal, P., Pupier, R., Richoux, P., Allemand, R., Genest, L.C. and Clary, J. (2000) *Coléoptères de Rhône-Alpes: Carabiques et Cicindèles*. Muséum d'Histoire Naturelle de Lyon et Société Linnéenne de Lyon, Lyon, France.

Desender, K. (1986) *Distribution and Ecology of Carabid Beetles in Belgium (Coleoptera, Carabidae), Part 1–4*. Institut Royal des Sciences Naturelles de Belgique, Brussels.

Duchesne, L.C., Lautenschlager, R.A. and Bell, F.W. (1999) Effects of clear-cutting and plant competition control methods on carabid (*Coleoptera: Carabidae*) assemblages in northwestern Ontario. *Environmental Monitoring and Assessment* 56, 87–96.

Egler, F.E. (1954) Vegetation Science Concepts I. Initial floristic composition, a factor in old-field vegetation development. *Vegetatio* 4, 412–417.

Garnier, A. (1965) La forêt de Montargis. Excursion du 4 avril 1965. *Bulletin de l'Association des Naturalistes de l'Orléanais* 30, 12–23.

Helms, J.A. (ed.) (1998) *The Dictionary of Forestry*. The Society of American Foresters and CAB International, Wallingford, UK.

Hurka, K. (1996) *Carabidae of the Czech and Slovak Republics*. Kabourek, Zlin.

Jeannel, R. (1941) *Faune de France. Coléoptères Carabiques*. Office central de Faunistique, Paris.

Koch, K. (1989) *Die Käfer Mitteleuropas. Ökologie. Band E1: Carabidae-Micropeplidae*. Goecke and Evers, Krefeld, Germany.

Koivula, M. (2002) Alternative harvesting methods and boreal carabid beetles (*Coleoptera, Carabidae*). *Forest Ecology and Management* 167, 103–121.

Koivula, M., Kukkonen, J. and Niemela, J. (2002). Boreal carabid-beetle (*Coleoptera, Carabidae*) assemblages along the clear-cut originated succession gradient. *Biodiversity and Conservation* 11, 1269–1288.

Lindroth, C.H. (1974) *Handbooks for the Identification of British Insects. Coleoptera Carabidae*. Society of London, London.

Magura, T., Tothmeresz, B. and Elek, Z. (2003) Diversity and composition of carabids during a forestry cycle. *Biodiversity and Conservation* 12, 73–85.

Niemelä, J., Halme, E. and Haila Y. (1990) Balancing sampling effort in pitfall trapping of carabid beetles. *Entomologica Fennica* 1, 233–238.

Niemelä, J., Langor, D. and Spence, J.R. (1993) Effects of clear-cut harvesting on boreal ground-beetle assemblages (*Coleoptera: Carabidae*) in Western Canada. *Conservation Biology* 7, 551–561.

Oliver, C.D. and Larson, B.C. (1996) *Forest Stand Dynamics: Update Edition*. John Wiley & Sons, New York.

ONF (1996) *Révision D'Aménagement. Forêt Domaniale de Montargis (4090.42 ha). Département du Loiret. 1996–2015.* ONF, Orléans, France.

ONF (1997) *La futaie Régulière de Chêne en Région Centre.* ONF DR Centre, Orléans, France.

Pinheiro, J.C. and Bates, D.M. (2000) *Mixed-effects Models in S and S-PLUS.* Springer, New York.

Rykiel, E.J. (1985) Towards a definition of ecological disturbance. *Australian Journal of Ecology* 10, 361–365.

Thiele, H. (1977) *Carabid Beetles in Their Environments. A Study on Habitat Selection by Adaptation in Physiology and Behaviour.* Springer-Verlag, Berlin.

Third Ministerial Conference on the Protection of Forests in Europe (ed.) (1998) Follow-up reports on the Ministerial Conferences on the protection of Forest in Europe. vol. 2 Sustainable forest management in Europe. Special report on the follow-up on the implementation of resolutions H1 and H2 of the Helsinki Ministerial Conference. *Third Ministerial Conference on the Protection of Forests in Europe, Lisbon, 2–4/06/1998.* Ministry of Agriculture, Rural Development and Fisheries, Lisbon.

White, G.C. and Bennetts, R.E. (1996) Analysis of frequency count data using the negative binomial distribution. *Ecology* 77, 2549–2557.

Worthen, W.B. (1996) Community composition and nested-subset analyses: Basic descriptors for community ecology. *Oikos* 76, 417–426.

Changes in the Composition of Wytham Woods (Southern England) 1974–2002, in Stands of Different Origins and Past Treatment

14

K.J. Kirby

English Nature, Northminster House, Peterborough, UK

One hundred and sixty-three permanent quadrats were recorded in 1974, 1991 and 2001 in Wytham Woods, a mixed, 320 ha woodland in Oxfordshire, southern England. Data were collected on tree and shrub layers and ground flora changes. Many aspects of the woods have remained stable over the 27 years since the first recording: there has been little change in overall tree composition or in average ground flora richness. However, the shrub layer has declined; cover of *Rubus fruticosus* has declined, but that of *Brachypodium sylvaticum* has increased, probably because of increased deer browsing. Differences in the composition of ancient and recent semi-natural stands declined over the period; some ancient woodland species spread further into recent woodland (e.g. *Carex pendula*).

14.1 Introduction

Peterken (1974, 1977) proposed that woods that were ancient (site wooded since at least AD 1600) and semi-natural (composed mainly of native trees and shrubs that have not obviously been planted) were of highest concern for nature conservation in the UK, because they were more likely to contain specialist woodland plants and animals. Such woods have since been listed in the ancient woodland inventories (Spencer and Kirby, 1992). Work elsewhere (e.g. Hermy *et al.*, 1993; Honnay *et al.*, 1999) has shown similar differences in the distribution of plant species according to the history of the site. Much of the

©CAB International 2004. *Forest Biodiversity: Lessons from History for Conservation*
(eds O. Honnay, K. Verheyen, B. Bossuyt and M. Hermy)

work comparing ancient and recent woodland has focused on isolated woods, which are predominantly either all ancient or all recent. However, many large woods consist of mixtures of both ancient and recent stands, with some areas that are plantations interspersed with the more natural stands.

This study examines the differences in the structure and composition of such a mixed site, Wytham Woods, near Oxford, in southern England, using data from permanent plots that have been recorded on three occasions between 1974 and 2002.

14.2 Site

Wytham Woods (National Grid Reference SP4608; 51°41′N, 1°19′W) is a mosaic of ancient and recent woodland, semi-natural stands and plantations of various ages and species, with small areas of open grassland and scrub (Elton, 1966; Perrins, 1989). Ancient and recent woodland were distinguished from historical data and field studies (Grayson and Jones, 1955; Gibson, 1986). Plantations were defined from forestry stock maps and field observations. The ancient semi-natural areas (36% of the 320 ha studied) were mainly former coppice; the recent semi-natural stands (31%) have grown up on former pasture over the past 100 years; most of the plantations were established in the past 50 years, either within the existing ancient woodland (12%) or on former open grassland (21%) (Fig. 14.1).

The woods lie on a hill with shallow calcareous soils over corallian limestone on the top and heavier clay soils on the lower slopes. The ancient

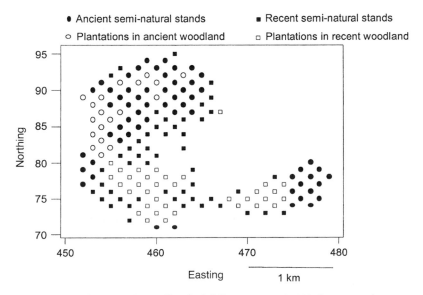

Fig. 14.1.   Distribution of woodland of different types in Wytham woods.

woodland is mainly on the lower slopes. The main tree species are *Quercus robur, Fraxinus excelsior, Acer pseudoplatanus*, with lesser amounts of *Fagus sylvatica* – even the plantations are predominantly broadleaved (Kirby *et al.*, 1996). The woods fall mainly into the *Fraxinus excelsior–Acer campestre–Mercurialis perennis*-type in the national vegetation classification (Rodwell, 1991).

14.3 Methods

Between 1973 and 1976 (the '1974' survey) 164 quadrats, each 10×10 m, were established and recorded at alternate points on a 100×100 m grid across the wood (Dawkins and Field, 1978). Each quadrat was offset 14.1 m to the north-east (true bearing) from grid-intersection marker posts; metal markers were buried at the south-west and north-east corners of the quadrats to assist with their relocation. In 1991/1992 (the '1991' survey), 163 of the quadrats were re-recorded and a further recording was made between 1999 and 2002 (the '2001' survey). In the 2001 survey about 90% of the quadrats were relocated precisely (based on finding the underground markers); where these were not found again the quadrats were relocated to within a few metres of the original positions based on the position of features – such as paths or measured trees – recorded in the 1974 survey. Each re-recording took place between April and August, but each quadrat was visited only once in each resurvey.

The recording methods are based on those used by Dawkins and Field (1978), with only minor modifications (Kirby *et al.*, 1996; Kirby and Thomas, 2000). A tape was laid across the south-west to north-east diagonal of the plot and percentage vegetation cover immediately above the tape was estimated in three height bands: canopy cover > 2.5 m high; shrub cover 0.5–2.5 m high; and ground cover < 0.5 m high. The canopy cover component was then split by tree species. Data on the largest trees in the quadrat and on the basal area of the trees in the surrounding woodland were collected but are not used in this chapter.

All vascular plants within the plot were identified, although in the ground flora analysis tree and shrub species have been excluded. Thirteen circular subplots, spaced equally across the two diagonals, each 0.1 m², were recorded. Not all species in the quadrat occurred within these circlets; therefore an index of abundance using the circlet results was produced by adding 1 to the number of times a species was recorded from the circlets for all species that occurred in the quadrat. Changes in the abundance of selected species using this index were calculated for 1974, 1991 and 2001 for the quadrats in which the species occurred. The species chosen were *Rubus fruticosus* and *Brachypodium sylvaticum*, species that appear to have responded negatively and positively, respectively, to changing deer pressure in the woods; *Urtica dioica*, which might be expected to respond to changing nutrient inputs; *Chamerion angustifolium*, a species common in disturbed conditions, often associated with plantations; and *Hyacinthoides non-scripta* a widespread ancient woodland species in the woods.

Additionally, cover estimates were made in 1974 for six common species with contrasting ecological requirements (*C. angustifolium*, *H. non-scripta*, *M. perennis*, *Pteridium aquilinum*, *R. fruticosus* and *U. dioica*), and in 1991 and 2001 cover estimates were made for all ground flora species. These cover data are not used in this chapter.

Where there was uncertainty as to whether species had been recorded consistently across the three recording periods, species were combined prior to the analysis. Species were classified as either 'ancient woodland species' for southern England (Marren, 1990); other species associated with woodland in Britain (though not confined to this habitat) (Kirby, 1988); or non-woodland species, mainly those more strongly associated with grassland. The class to which a species was allocated was defined independently of the Wytham surveys, and these classes have been used widely in conservation evaluations over the past 20 years (Anon., 1989).

Comparisons were made across the three surveys in terms of the composition and cover of different woody species for stands of different origins and treatments. Differences in species richness in different types of woodland were compared, in particular the differences in distribution of ancient woodland species between ancient and recent semi-natural stands; and between the flora of plantations and semi-natural stands. Species names follow Stace (1991).

14.4 Results

Over the wood as a whole the canopy became more open between 1974 and 1991, but closed again between 1991 and 2001 (Table 14.1). Plantations tended to have denser canopy cover than semi-natural areas.

The composition of the tree and shrub layer, as estimated by the canopy cover, remained stable, with ash (*F. excelsior*), sycamore (*A. pseudoplatanus*) and oak (*Q. robur*) forming the bulk of the cover. Beech (*F. sylvatica*) was a significant component of the woods except in the ancient semi-natural stands; birch (*Betula* spp.) was common in the ancient semi-natural areas in 1974, but declined. Conifers (such as spruce, *Picea abies*) remained a small part of the plantation area (Table 14.2). The shrub layer declined greatly since 1974 across all the wood.

Ground flora cover declined between 1974 and 1991, but there was little change thereafter across the wood. The semi-natural stands showed little change in species richness over the period, but under plantations on both ancient and recent sites the mean number of species per plot declined between 1974 and 2001.

Table 14.1. Composition of the tree layer in 1974, 1991 and 2001 based on percentage cover estimates across the south-west/north-east diagonal of each quadrat.

	WW	ASN	RSN	AP	RP
Number of quadrats	163	59	50	19	35
Overall canopy cover					
1974	81 (2)	84 (3)	80 (4)	72 (7)	83 (4)
1991	69 (2)	68 (4)	57 (5)	75 (6)	83 (4)
2001	77 (2)	76 (3)	70 (3)	89 (3)	83 (4)
Ash (*Fraxinus excelsior*)					
1974	18 (2)	17 (4)	18 (4)	22 (7)	19 (6)
1991	19 (2)	17 (3)	19 (4)	21 (7)	22 (6)
2001	23 (2)	26 (4)	27 (4)	29 (7)	17 (5)
Oak (*Quercus robur*)					
1974	12 (2)	14 (4)	7 (3)	14 (4)	14 (4)
1991	10 (1)	11 (2)	6 (2)	11 (3)	12 (4)
2001	9 (2)	11 (3)	6 (2)	13 (6)	11 (4)
Sycamore (*Acer pseudoplatanus*)					
1974	18 (2)	20 (4)	22 (5)	12 (6)	10 (4)
1991	15 (2)	16 (3)	15 (4)	18 (8)	11 (4)
2001	17 (2)	21 (4)	17 (4)	20 (7)	8 (4)
Beech (*Fagus sylvatica*)					
1974	10 (2)	0	12 (4)	15 (8)	24 (5)
1991	10 (2)	0	9 (4)	17 (8)	24 (6)
2001	12 (2)	0	11 (4)	19 (9)	31 (6)
Birch (*Betula* spp.)					
1974	5 (1)	9 (3)	3 (2)	3 (2)	1 (1)
1991	3 (1)	6 (2)	1 (1)	1 (1)	2 (1)
2001	2 (1)	4 (2)	0	1 (1)	2 (1)
Conifers					
1974	2 (1)	0	0	3 (1)	9 (2)
1991	3 (1)	0	0	3 (2)	11 (3)
2001	3 (1)	0	0	4 (3)	4 (3)
Shrub cover					
1974	44 (3)	41 (4)	41 (5)	56 (8)	45 (6)
1991	24 (1)	30 (7)	25 (3)	25 (5)	14 (3)
2001	17 (1)	17 (2)	19 (3)	10 (3)	6 (2)

WW, Whole wood; ASN, ancient semi-natural; RSN, recent semi-natural; AP, plantations in ancient woodland; RP, plantations in recent woodland.
Standard errors are shown in parentheses.

In 1974 the mean number of ancient woodland species was greater in ancient semi-natural areas than in recent semi-natural stands; by 2001 this difference had gone. Some ancient woodland species were still more common in ancient semi-natural woodland: for example *Carex sylvatica*, *H. non-scripta*; others had increased in recent woodland (*C. pendula*); while some, such as

Table 14.2.   The flora of different sections of the woodland in 1974, 1991 and
2001.

	WW	ASN	RSN	AP	RP
Number of quadrats	163	59	50	19	35
Mean % ground cover					
1974	80 (2)	86 (3)	79 (3)	85 (5)	69 (6)
1991	64 (2)	72 (3)	65 (5)	56 (7)	54 (6)
2001	67 (3)	75 (3)	68 (5)	54 (8)	61 (6)
Mean number of					
species/quadrat					
1974	16.7 (0.3)	17.4 (0.7)	15.2 (1.4)	17.3 (1.6)	17.7 (1.6)
1991	17.2 (0.8)	17.8 (1.0)	17.5 (1.0)	17.7 (2.0)	15.2 (2.0)
2001	15.4 (0.6)	17.5 (0.9)	15.9 (1.3)	11.2 (1.4)	13.6 (1.3)
Mean ancient wood					
species					
1974	2.5 (0.1)	3.3 (0.3)	1.7 (0.2)	3.0 (0.4)	2.1 (0.3)
1991	2.6 (0.8)	2.6 (0.2)	2.8 (0.3)	2.6 (0.5)	2.6 (0.4)
2001	2.8 (0.2)	3.0 (0.2)	3.2 (0.4)	2.1 (0.4)	2.1 (0.3)

WW, Whole wood; ASN, ancient semi-natural; RSN, recent semi-natural; AP,
plantations in ancient woodland; RP, plantations in recent woodland.
Standard errors are shown in parentheses.

Festuca gigantea and *Tamus communis*, were not at all restricted to ancient
semi-natural woodland in Wytham (Table 14.3). By contrast, some widespread woodland species, such as *Ajuga reptans* and *P. aquilinum*, showed as
great a difference in their frequencies between ancient and recent stands as
some ancient woodland species. Within the plantations in ancient woodland,
low-growing herbs tended to be less common within replanted areas than
within the semi-natural stands (Table 14.4).

   R. fruticosus and *B. sylvaticum* illustrate the marked changes in the appearance of the woodland that have occurred since 1974, from one where *Rubus*
was the major ground flora species to the current grass-dominance over large
areas (Table 14.5). *U. dioica* shows no significant change in the cover index. *H.
non-scripta* shows little overall change, but more variability; some of this may
be due to differences in recording time, since its cover varies considerably
between April and August. *C. angustifolium* was relatively common in the
plantation quadrats in 1974, but was not recorded at all in 2001.

14.5 Discussion

Many aspects of the woods have remained stable over the 27 years since the
first recording: there has been little change in overall tree composition or in

Table 14.3.   Differences in occurrence of ancient woodland species between
ancient and recent semi-natural woodland in 1974 and 2001.

Species	1974 occurrences		2001 occurrences		Total occurrences	
	ASN	RSN	ASN	RSN	1974	2001
Number of quadrats	59	50	59	50	163	163
Number of ancient woodland species	25	15	21	22	27	28
Ancient woodland species						
Carex sylvatica	22	14	37	24	45	83
Hyacinthoides non-scripta	26	11	33	15	62	79
Carex pendula	22	6	20	13	43	54
Festuca gigantea	20	13	14	18	53	38
Lamiastrum galeobdolon	21	6	20	6	33	34
Carex remota	1	0	8	3	1	24
Veronica montana	10	2	13	5	13	23
Primula vulgaris	12	4	15	4	19	22
Tamus communis	11	13	5	7	35	17
Potentilla sterilis	3	4	7	4	12	15
Oxalis acetosella	5	1	8	3	7	13
Lysimachia nemorum	9	2	7	2	12	10
Ribes sylvestre	3	0	7	1	5	10
Holcus mollis	8	1	5	2	16	9
Bromopsis ramosa	3	5	0	0	13	2
Other ancient woodland species	18	10	13	7	43	30
Selected other woodland species						
Ajuga reptans	24	7	10	5	41	19
Dryopteris filix-mas	30	15	30	12	61	59
Lonicera periclymenum	17	5	4	2	33	7
Pteridium aquilinum	40	24	35	16	99	70

ASN, Ancient semi-natural; RSN, recent semi-natural.
Only ancient woodland species with at least 10 occurrences in either 1974 or 2001
are shown.

mean ground flora richness. The semi-natural stands have become more open
(partly through management – the widening of rides through the woods,
partly natural openings such as windblow), whereas growth of the plantations
(many of them still relatively young in 1974) has led to the maintenance of
closed canopies. The greater density of the canopy and the higher proportions
of beech and conifers in plantations compared to semi-natural stands are likely
to be the major factors in the lower species richness of plantation quadrats in
2001. Since there was less major forestry disturbance in the woods since 1974
compared to the period prior to 1974, *C. angustifolium* (present in 26 out of 54
plantation quadrats in 1974) became less common.

Table 14.4. Differences in the frequency of species in semi-natural and plantations stands in 2001.

	ASN	AP	RSN	RP
Total number of quadrats	59	19	50	35
Species				
Ajuga reptans	10	0	5	4
Cardamine flexuosa	16	2	3	5
*Festuca gigantea**	14	0	18	6
Geum urbanum	29	5	27	12
Glechoma hederacea	42	8	30	13
*Lamiastrum galeobdolon**	20	3	6	5
*Primula vulgaris**	15	2	4	1
Prunella vulgaris	11	1	8	2
Ranunculus repens	14	0	6	7
Rumex spp.	24	4	14	17
Veronica chamaedrys	28	2	10	5
*Veronica montana**	13	1	5	4
Viola riviniana	19	2	22	12

Species listed are those with a total occurrence of at least 10 quadrats in ancient woodland and a ratio of ancient semi-natural to plantation in ancient woodland occurrences of 5 : 1.
ASN, Ancient semi-natural; RSN, recent semi-natural; AP, plantations in ancient woodland; RP plantations in recent woodland.
*Ancient woodland species.

There was a major increase in deer numbers in the wood between 1974 and 1991; since then, various control measures have been introduced. The increased browsing led to loss of the understorey, as seen in the decline in the shrub layer and cover of *R. fruticosus*, and increase in grasses such as *B. sylvaticum*. Kirby and Thomas (2000) found no evidence for eutrophication effects on the ground flora – the wood is large and hence the quadrats well buffered from changes in surrounding land. *U. dioica*, which might be expected to respond to any increased nutrients, shows no sign of increasing its cover.

Ancient and recent semi-natural woodland at Wytham does differ, but not all proposed ancient woodland species show a difference: several are common across the recent woodland (e.g. *F. gigantea*). The differences between the ancient and recent sections of the wood seem to have reduced somewhat since 1974, partly because of spread of species into recent quadrats – for example *C. pendula* (from 6 to 13 occurrences). This species is not very palatable to deer and has spread in woods elsewhere under high deer impacts (Crampton *et al.*, 1998). Its seed could easily be moved in soil on deer hooves, while removal of taller, palatable competitor species such as *R. fruticosus* may also have contributed to its success.

Table 14.5. Changes in the mean cover index (out of 14) for quadrats in which a species occurred, for selected species 1974–2001.

Species	Whole wood		Cover index for quadrats where species occurred			
	WWF	WWA	ASN	RSN	AP	RP
Rubus fruticosus						
1974	142	4.8 (0.2)	4.6 (0.4)	4.6 (0.5)	6.7 (0.6)	4.5 (0.6)
1991	127	3.5 (0.3)	3.3 (0.4)	4.7 (0.6)	2.8 (0.5)	2.6 (0.3)
2001	127	2.5 (0.2)	2.4 (0.2)	3.2 (0.4)	1.7 (0.2)	2.2 (0.3)
Brachypodium sylvaticum						
1974	63	2.3 (0.3)	2.1 (0.3)	2.7 (0.6)	3.0 (1.3)	1.8 (0.4)
1991	136	4.7 (0.3)	4.5 (0.4)	4.8 (0.6)	3.7 (1.1)	5.6 (0.9)
2001	142	5.8 (0.3)	5.8 (0.5)	6.1 (0.6)	4.6 (1.1)	6.1 (0.8)
Urtica dioica						
1974	122	4.0 (0.3)	4.8 (0.5)	3.7 (0.6)	4.1 (0.9)	2.6 (0.4)
1991	129	3.6 (0.3)	4.2 (0.4)	3.1 (0.4)	3.3 (1.0)	3.4 (0.5)
2001	121	3.5 (0.3)	4.0 (0.4)	2.9 (0.5)	3.2 (1.0)	3.3 (0.6)
Hyacinthoides non-scripta						
1974	62	3.0 (0.3)	2.6 (0.4)	3.0 (0.8)	5.6 (1.0)	1.9 (0.4)
1991	63	3.2 (0.4)	2.9 (0.6)	2.3 (0.9)	4.3 (1.0)	3.5 (0.9)
2001	79	2.9 (0.4)	2.8 (0.6)	1.6 (0.5)	3.6 (1.0)	3.8 (0.8)
Chamerion angustifolium			For this species the cover index was so low that simply the number of occurrences have been given for the different sections of the wood.			
1974	44	1.7 (0.3)	9	9	9	17
1991	27	0.6 (0.1)	4	4	8	11
2001	0	0	0	0	0	0

WWF, Number of quadrats in which the species occurred; WWA, mean cover index for whole wood; ASN, cover index for ancient semi-natural occurrences; RSN, cover index for recent semi-natural occurrences; AP, cover index for ancient plantation occurrences; RP, cover index for recent plantation occurrences. Values in brackets are standard errors.

A blurring of the distinction between ancient and recent woodland, with respect to the occurrence of ancient woodland species, would be expected at a site such as Wytham because the recent semi-natural stands are in close proximity to the ancient woodland (Peterken and Game, 1984; Brunet and Oheimb, 1998; Bossuyt *et al.*, 1999). In addition, the recent stands are, in some cases, nearly 200 years old, and they may incorporate small fragments of ancient woodland from the former landscape that were not picked up on the maps used to define ancient woodland for this study. Nevertheless, some differences were still found in the abundance of various ancient woodland species (e.g. *H. non-scripta*, *Primula vulgaris*), but also of widespread species (e.g. *A. reptans*, *Dryopteris filix-mas*). These are not likely to be dispersal-limited, so the greater abundance in ancient stands is more likely to be the result of the difference in the soils underlying ancient and recent parts of Wytham Woods. If this factor is not allowed for in plant distribution studies, then misleading conclusions might be reached as to whether species are ancient woodland species or not. The preference of a species for ancient woodland may also be masked by the effects of plantations: *Lamiastrum galeobdolon*, *P. vulgaris* and *Veronica montana* all show a greater frequency in ancient stands if the semi-natural areas are compared, but the general reduction in species abundance in the plantations means that no difference could be detected.

The data have shown the value of the original plot system established by Dawkins and Field (1978) in allowing us to track the subsequent changes to the woods, including some that were not predictable (the massive expansion in deer populations), or not even known about (ancient woodland species distributions) when they started work on the system in 1973.

Acknowledgements

I would like to thank Oxford University for permission to work in the woods; all those who have helped with the recording over the years, particularly Rachel Thomas, Jeanette Hall, Hannah Isles, Emma Goldberg and Rebecca Isted. However, without the efforts and inspiration of Colyear Dawkins and David Field in setting the system up and undertaking the 1974 recording, none of the subsequent studies would have been possible.

References

Anon. (1989) *Guidelines for the Selection of Biological Sites of Special Scientific Interest*. Nature Conservancy Council, Peterborough, UK.

Bossuyt, B., Hermy, M. and Deckers, J. (1999) Migration of herbaceous plant species across ancient–recent forest ecotones in central Belgium. *Journal of Ecology* 87, 628–638.

Brunet, J. and von Oheimb, G. (1998) Migration of vascular plants to secondary woodlands in southern Sweden. *Journal of Ecology* 86, 429–438.

Crampton, A.B., Stutter, O., Kirby, K.J. and Welch, R.C. (1998) Changes in the composition of Monks Wood National Nature Reserve (Cambridgeshire) 1964–1996. *Arboricultural Journal* 22, 229–245.

Dawkins, H.C.D. and Field, D.R.B. (1978) *A Long-Term Surveillance System for British Woodland Vegetation*. Commonwealth Forestry Institute, Oxford, UK.

Elton, C.S. (1966) *The Pattern of Animal Communities*. Methuen, London.

Gibson, C.W.D. (1986) Management history in relation to changes in the flora of different habitats on an Oxfordshire estate, England. *Biological Conservation* 38, 217–232.

Grayson, A.J. and Jones, E.W. (1955) *Notes On The History of the Wytham Estate with Special Reference to the Woodlands*. Imperial Forestry Institute, Oxford, UK.

Hermy, M., van den Bremt, P. and Tack, G. (1993) Effects of site history on woodland vegetation. In: Broekmeyer, M.E.A., Vos, W. and Koop, H. (eds) *European Forest Reserves*. Pudoc, Wageningen, The Netherlands, pp. 219–232.

Honnay, O., Hermy, M. and Coppin, P. (1999) Effects of age and diversity of forest patches in Belgium on plant species richness and implications for conservation and reforestation. *Biological Conservation* 87, 73–84.

Kirby, K.J. (1988) *A Woodland Survey Handbook*. Nature Conservancy Council, Peterborough, UK.

Kirby, K.J. and Thomas, R.C. (2000) Changes in the ground flora in Wytham Woods, southern England from 1974 to 1991 – implications for nature conservation. *Journal of Vegetation Science* 11, 871–880.

Kirby, K.J., Thomas, R.C. and Dawkins, H.C. (1996) Monitoring of changes in tree and shrub layers in Wytham Woods (Oxfordshire), 1974–1991. *Forestry* 69, 319–334.

Marren, P. (1990) *Woodland Heritage*. David and Charles, Newton Abbot, UK.

Perrins, C.M. (1989) Wytham Woods. *Biologist* 36, 5–9.

Peterken, G.F. (1974) A method for assessing woodland flora using indicator species. *Biological Conservation* 6, 239–245.

Peterken, G.F. (1977) Habitat conservation priorities in British and European woodland. *Biological Conservation* 11, 223–236.

Peterken, G.F. and Game, M. (1984) Historical factors affecting the number and distribution of vascular plant species in the woodlands of central Lincolnshire. *Journal of Ecology* 72, 155–182.

Rodwell, J. 1991 *British Plant Communities: 1 Woodlands and Scrub*. Cambridge University Press, Cambridge, UK.

Spencer, J.W. and Kirby, K.J. (1992) An inventory of ancient woodland for England and Wales. *Biological Conservation* 62, 77–93.

Stace, C. (1991) *A New Flora of the British Isles*. Cambridge University Press, Cambridge, UK.

Forest History, Continuity and Dynamic Naturalness

15

C. Westphal, W. Härdtle and G. von Oheimb

Institute of Ecology, University of Lüneburg, Lüneburg, Germany

Forest history is one of the bases by which biodiversity in woodlands can be explained. In practice, forest history is mainly investigated in relation to research about ancient woodlands, or to particular scientific questions. However, it seems necessary to assess forest history not only for specific occasions but to take it as a base for an evaluation of the naturalness of any particular forest. Naturalness is a widely used criterion for evaluating the current state of a forest in relation to specific nature conservation values. It refers to the actual composition of the vegetation, neglecting any historical processes. To assess naturalness within a dynamic concept means to display a set of criteria that reflect the individual historical development of a forest. This has to include certain key processes, such as the continuity of woodland existence, the process of regeneration, competition, ageing and dying, but also a set of criteria characterizing the human interference during all the reconstructable past.

15.1 Introduction

Naturalness is a widely used criterion for evaluating the nature conservation value of a landscape (Usher and Erz, 1994). Within forestry, naturalness is one of the key elements used in biotope mapping (Ammer and Utschick, 1982; Hanstein and Sturm, 1986; Hofmann and Jenssen, 1999) and in the broad-scale evaluation of woodland as landscape (Grabherr *et al.*, 1996), and may sometimes even be assessed in the course of a forest inventory (Michiels, 1999).

©CAB International 2004. *Forest Biodiversity: Lessons from History for Conservation*
(eds O. Honnay, K. Verheyen, B. Bossuyt and M. Hermy)

The method of assessment is theoretically based either on the concept of the potential natural vegetation (Tüxen, 1956) or on the concept of hemeroby (Jalas, 1955; Sukopp, 1976), or on a mixture of both. However, taking into account the woodland ecology concepts published or discussed in the past decade, it seems that it is necessary to integrate the dynamic aspects of nature as well as forest history into the concept of naturalness.

15.2 Static Concepts for the Evaluation of Naturalness In Forests

The evaluation of naturalness in forests has a theoretical basis which is either the concept of potential natural vegetation (PNV) or the concept of hemeroby. Both concepts impose a static view on nature because they focus exclusively on the current state of a forest (i.e. the species composition) without considering the historical dynamic that has produced it.

15.2.1 The concept of the potential natural vegetation (PNV), from Tüxen (1956)

The concept of the potential natural vegetation, which is widely used nowadays, was first described by Tüxen in 1956 as a method of determining the present potential of a site in terms of its natural vegetation cover. Following this concept, the evaluation of naturalness compares the present state of a forest stand (i.e. its composition of tree species) with the desired species composition of the hypothetically constructed PNV. Naturalness, usually described in steps or classes, is determined by the similarity between the two states. This concept has given rise to a lively scientific debate about the construction mode for the PNV (e.g. Neuhäusl, 1984; Kowarik, 1987; Fischer, 1995; Härdtle, 1995; Schmidt, 1998; Reif, 1999/2000). Neophytic tree species such as the Douglas fir (*Pseudotsuga menziesii* [Mirbel] Franco) are considered to be part of the PNV, as they can compete successfully with the natural species in different site-types (Knoerzer, 1999).

15.2.2 The concept of hemeroby, from Jalas (1955) and Sukopp (1976)

The concept of hemeroby is an actualistic concept, with no relationship to historical development. It focuses on the present distribution of species at sites differently affected by human impact: 'Hemeroby is understood as the sum of effects that take place by planned and unplanned human interference in ecosystems. The stage of hemeroby results from effects on the site and its organisms' (Sukopp, 1976, p. 21).

15.3 Theoretical Criticisms of the Static Approach

The concept of the PNV which focuses on the climax stage has been criticized for being static and for not taking into account the different successional stages (Schmidt, 1998). Reif (1999/2000) has therefore demanded a dynamic expansion of the concept to include all tree species and succession phases that belong to the natural vegetation dynamics. He also suggested the integration of such elements as dead wood and structural diversity to complete the 'reference image' of a natural forest as the basis for the evaluation of naturalness.

But even with such an expansion, two major problems connected with any construction or prediction of a reference state as a basis of evaluation remain unsolved:

1. When constructing a potential natural state, *representative values* for various ecological measures (e.g. the natural amount of dead wood, the distribution of successional stages, the percentages of tree species in the successional mosaic, typical soil properties) have to be given. This is highly problematic because of the lack of existing natural forests in nearly all the forest communities of Central Europe. It seems absolutely impossible to predict the mixture of tree species, the amount of dead wood and the successional mosaic, because there are no surviving examples of primary woodland left as a basis for scientific orientation. In addition, these features vary in nature and are time-dependent. It is therefore a task for forest ecology to investigate these questions, but present knowledge is far too limited to predict the exact values neccessary for the definition of a natural state.

2. Restricting the definition to the present state, even if it is defined by a variety of additional ecological factors (criteria), *neglects the historical processes* and the dynamics that underlie the current *status quo*. A natural composition of tree species can be the result of a plantation with intensive soil preparation, fertilization and repeated heavy thinning, but can also result from a natural regeneration process over several decades without further human interference. To exclude the historical processes and to focus solely on the present state can result in serious shortcomings in the interpretation of naturalness as a whole.

In recent decades there has been a broad-based discussion about forest ecology and the goals of nature protection in forests (e.g. Pickett and White, 1985; Ellenberg, 1987; Bundesforschungsanstalt für Naturschutz und Landschaftsökologie, 1989; Remmert, 1991; Pickett *et al.*, 1992; Otto, 1993, 1994, 1995; Sturm, 1993, 1994; Niedersächsische Landesforstverwaltung, 1994; Plachter, 1996; Scherzinger, 1996; Jedicke, 1998; Jax, 1998/1999; Perpeet, 1999). In searching for a common thread running through all these publications, a theme that emerges is that forests are understood much more as dynamic entities, emphasizing the function of natural processes. Forests are seen as a 'multivariable stochastic successional mosaic' (Sturm, 1994, p. 14).

They are composed of biotic and abiotic matter which is driven by various dynamic forces and a range of different processes. The quantity and quality of the matter can be seen as a temporary result of these processes. The role of man and the human impact on the present state of nature has become increasingly significant scientifically (Nilsson and Grelsson, 1995; Plachter, 1995; Peterken, 1996; Schneider and Poschlod, 1999).

15.4 An Expanded Definition of Naturalness in Forests

The following definition of naturalness relates specifically to forests and is not directly applicable to other formations. An expansion to incorporate the dynamic elements does not necessarily mean excluding the static elements. As nature generally is observed in a static state, which is the temporary result of constant dynamic change, both elements have to be part of an expanded definition. In this sense, naturalness can be defined as follows:

> Naturalness in forests is a process-related measure that develops in relation to time span as well as to intensity of anthropogenically undisturbed woodland dynamics. Its material expression is found in a continuously changing set of species and structures resulting from these dynamics.

Under this definition, naturalness is a measure that cannot reach a maximum but – in anthropogenically undisturbed conditions – is constantly increasing. It cannot be created artificially or accelerated as it results from undisturbed dynamics tied to real time. Its magnitude depends on the time and intensity of anthropogenically undisturbed development. The longer the natural processes evolve, the more naturalness can accumulate. Natural processes also take place in anthropogenically modified forests as soon as that impact ceases. Species and structures are to be seen as a result of these processes, with a temporary existence. An expanded evaluation of naturalness must therefore include both static and dynamic elements. Table 15.1 summarizes the major differences between the static and dynamic approaches to naturalness.

In so far as the past can be reconstructed, an assessment of naturalness must describe how extensively the natural processes have been changed. It should cover all the important processes by setting up a number of adequate criteria. However, to put this into practice raises a series of practical questions.

15.5 Proposals for a Dynamic Extension of the Concept of Naturalness

The main characteristic of dynamic naturalness is the focus on the historical development of a forest. The basis of the evaluation does not have to be hypothetically constructed (as has to be the PNV), but can be reconstructed from

Table 15.1. Comparison of the static and the dynamic evaluation approaches to naturalness.

	Static approach	Dynamic approach
Basis of evaluation	Hypothetical construction of an assumed natural state	Reconstruction of a true historical development
Focus of interest	Distribution or quantities of various ecological components or compartments (e.g. species, structures)	Historical development and human impact on specific woodland processes
Evaluation criteria	Result-related Quantifying	Process-related Prescribing
Aims of evaluation	Manipulate the present state (tree species) in accordance with the PNV	Explain the present state (species, structures) in relation to its history, in order to decrease future impacts

historical facts. An effort must be made to obtain appropriate information about the type and amount of human interference in the reconstructable past. This search for historical information can be time-consuming, but it is a single, non-recurring activity that, once done, has only to be continued. So far as current forestry practice is concerned, it requires that all silvicultural activities at the stand level are expressly recorded.

The individual historical development of a particular woodland has to be studied, focusing on the different ecological processes. Examples are given by Schaal (1994), Schneider and Poschlod (1999) and Westphal (2001). Certain key processes have to be distinguished that have a major influence on other biotic and abiotic components and thereby on the naturalness of the whole system. These processes are described by a set of general criteria comprising the major segments of a woodland life-cycle: woodland formation, woodland growth, woodland ageing. They are supplemented by a specific set of further criteria such as biomass accumulation, soil development and population dynamics of key faunistic species, such as large herbivores (Table 15.2). This set has to be seen as a first attempt to cover the relevant processes of woodland dynamics. Static criteria, such as the tree species composition (PNV), have to be added. The set has to be expanded and completed gradually.

Below, the criteria listed in Table 15.2 are described briefly.

15.5.1 Criteria of dynamic naturalness

The criteria of naturalness refer to key processes that are supposed to underly the current state of a woodland and which have led to its present appearance.

Table 15.2. Elements for an evaluation of naturalness: criteria and indicators of human impacts.

Criteria of naturalness	Indicators of human impact
• Continuity of woodland existence	
Continuity of forest land-use	Continuous forest land-use (allowing clear-cut)
Continuity of forest cover	Time span of continuous forest cover (excluding clear-cut)
Continuity of dead wood supply	Time span of continuous dead wood supply
• Woodland formation	
Spontaneity of natural regeneration	Regeneration type
Time span of natural regeneration	Time span of regeneration
• Woodland growth	
Spontaneity of tree competition	Intensity of pruning and thinning
Allowance of natural disturbances	Intensity of clearance or utilization of stands with natural disturbances
• Woodland ageing	
Age structure	Distribution of age classes or succession phases
Phases of ageing and decay	Age/DBH of trees at harvest
	Time span of harvest
	Area under strict protection
• Biomass accumulation	Average growing stock
• Soil development	
Chemical impacts on soil development	Direct input of chalk, fertilizer, pesticides
	Atmospheric emissions
Physical impacts on soil development	Soil cultivation type
	Other types of soil disturbance
• Population dynamics of deer	Grazing of naturally regenerating forest trees
• Tree species composition	Present tree species composition

However, their assessment focuses on the description of the nature and extent of the historical human impact. The criteria can refer either to factual impacts (e.g. soil cultivation, application of chemicals, clear-felling) or to present-state variables (e.g. biomass, age structure, mosaic distribution) which result from factual impacts that cannot be reconstructed directly, due to a lack of adequate information.

Continuity of woodland existence

Continuity is a criterion of increasing significance within forest ecology. Originally discussed within the context of ancient woodlands (Rackham,

1980; Peterken, 1981; Watkins, 1990), it has gained in scientific importance, bryology (Koperski, 1998), lichenology (Rose, 1986), mycology (Kost, 1989), soil ecology (Wilson *et al.*, 1997) and entomology (Aßmann, 1998). It has proved to be one of the major reasons for high nature conservation values and as a precondition for the existence of stenotopic indicator species which depend on very specific and typical forest habitats. Three subcriteria can be set out which reflect an increasing quality of continuity (Table 15.3).

Continuity of forest land-use

Continuity of forest land-use comes with constant forest use. Therefore the historical land use has to be assessed. Ancient and recent woodlands have to be distinguished, and the time span of constant forest use has to be recorded at the stand level. The continuity of forest land-use is a precondition for the existence, survival or recolonization of many autochthonous vascular plant species (Brunet and Von Oheimb, 1998; Brunet *et al.*, 2000). Ancient woodlands are woodlands that have been in existence continuously for at least several hundred years (Peterken, 1981; Norddeutsche Naturschutzakademie, 1994) but are allowed to be partially clearcut as long as they are re-afforested or allowed to regenerate immediately.

Continuity of forest cover

The continuity of forest cover focuses on tree species and their function as a habitat for xylobiont species. Continuity is ensured as long as a minimum growing stock is maintained that provides a forest climate that assures the survival of forest species. Clear-cut is regarded as a break in continuity.

Table 15.3.   Subcriteria for the continuity of woodland existence.

	Continuity of woodland existence		
	Continuity of forest land-use	Continuity of forest cover	Continuity of dead wood supply
Characteristic	Continuous forest use	Continuous forest cover (minimum growing stock)	Continuous existence of ageing and dying trees
Excluding	Other forms of land use	Clear-cut	Clear-cut and intensive utilization
Including	Clear-cut and intensive utilization	Intensive utilization	Extensive utilization

According to Rose (1986) and Fritz and Larsson (1996), this definition of continuity is relevant for xylobiont mosses and lichens and, within the category of ancient woodlands, differentiates between those that underwent clear-cut and those that did not. The minimum growing stock maintaining continuity has to be specified for the main forest associations.

Continuity of dead wood supply

This category represents the highest level of continuity. It comprises the time span during which not only forest cover but also dying and dead wood has been in existence. Continuity in the supply of dead wood is maintained by the absence of, or a limited amount of, utilization (Karström, 1992), or by a minimum amount of dead wood being allowed to remain on the site. In the latter case, continuity is a mirror of past and current silvicultural practices and reflects the intensity of utilization. The impact assessment is expected to discover areas of high continuity that have remained unnoticed so far, and thereby to bring them under protection.

Woodland formation

The criterion of woodland formation refers to the type of regeneration for a given stand. It can differ in many aspects from the natural regeneration process, which depends on factors such as the forest association, the occurrence of natural disturbances, the seed supply, the climatic conditions in most years, and other conditions driven by chance. In a managed forest the regeneration process is often shortened, as it begins with premature stands and is manipulated in various ways (e.g. spacing, species composition, interspecific competition). Regeneration is a process responsible for many aspects of structural diversity and the diversity of age classes (Mlinsek, 1967; Smejkal *et al.*, 1997). Two subcriteria can be distinguished.

Spontaneity of the regeneration process

The spontaneity of the regeneration process is highest in natural regeneration. It is responsible for the building of successional patches and mosaic structures (Koop, 1982; Koop and Hilgen, 1987; Tabaku and Meyer, 1999). The impact assessment has to specify the management activities during the re-afforestation period. Basically, a distinction has to be made between planting, replanting bare patches within natural regeneration, shelterwood systems, seeding and an undisturbed natural regeneration process.

Time span of the regeneration process

Another aspect of the regeneration process concerns the time span needed to complete regeneration and to regard a stand or a distinct area as being fully renewed. In primary woodland, natural regeneration is a continuous and endless process, whereas in managed forests time spans can vary between 1 and 2 years and up to several decades (Mayer, 1984).

Woodland growth

The criterion of woodland growth refers to all the processes following successful regeneration until the onset of ageing and decay. During this period the natural dynamic is formed by a variety of spontaneous processes. Management can alter these processes in many ways. The assessment of woodland growth focuses on two aspects.

Spontaneity of tree competition

The spontaneity of tree competition refers to all the processes of inter- and intraspecific competition among trees. It can be seen as a directed process (Jenssen and Hofmann, 1996) which results in a site-adapted species composition, a natural species diversity and a natural diversity of spatial structures. It can be altered by silvicultural measures such as intensive thinning and species regulation. The historical impact assessment has to focus on these measures and to record the type and intensity at the stand level.

Allowance of natural disturbances

During the life cycle of a woodland, a series of natural disturbances of different types and intensities occur and reset the direction of the natural dynamic. According to Jax (1998/1999), natural disturbances are characterized by their frequency, predictability, intensity, and spatial and temporal extension. They may create unpredictable turning points and irreversible changes in the dynamic, and are seen as one major source of the natural diversity of a woodland (Pickett *et al.*, 1992; Sturm, 1993; Jax, 1998/1999). However, management interferes profoundly with natural disturbances and thereby changes the dynamics. If and to what extent current silvicultural practices allow natural disturbances to exercise their effects on woodland dynamics have to be evaluated.

Woodland ageing

Woodland ageing begins in the second half of the life cycle, with the onset of physical weakening. At the stand level, these are the phases of ageing and decay, which are roughly estimated to cover 40–60% of the area within beech forest associations (Koop and Hilgen, 1987; Korpel, 1995; Tabaku, 1999). The vast majority of xylobiont species depend on the continuous existence of the specific structures that occur in these successional phases (Rose, 1986; Kost, 1989; Aßmann, 1998; Koperski, 1998). It is obvious that the ageing processes, which can last more than 200 years (Mlinsek, 1967), are almost absent in managed forests. The assessment of ageing processes thus has to focus on two different aspects.

Age structure

The age structure is a static parameter that shows the actual distribution of age classes. If the forest is already in a more natural state, it refers to the distribution of successional mosaics. This static criterion specifies a certain aspect of the development conditions of past woodland dynamics. Forests with an unbalanced age structure – broadly considered – cannot derive from the natural dynamics and cannot encompass the full range of natural processes. The appropriate information about the age structure can often be taken directly from the forest inventory or from biotope mapping.

Phases of ageing and decay

The phases of ageing and decay are directly connected with the age structure. An overview of the occurrence and proportion of these phases can be obtained by assessing an area which is under strict protection and is therefore dedicated to a natural process of decay. As regards current silvicultural practices, it has to evaluate harvesting operations in terms of the age, or diameter at breast height (DBH), of the harvested trees.

Biomass accumulation

This criterion is directed at natural forest associations and their site-specific biomass per hectare (volume of the standing crop) (Mayer, 1987; Smejkal *et al.*, 1995; Hofmann, 1997). The correspondence between the growing stock and the natural biomass is an indicator of the historical impact and reflects the maturity of the forest ecosystem. Information on the growing stock can be calculated from the data of the forest inventory.

Soil development

The soil development is a long-term dynamic resulting in different soil types and their specific characteristics. Due to the massive historical impact on forest soils, the scientific and ecological significance of relatively undisturbed soil is regarded as considerable (Ball and Stevens, 1981; Pruza, 1985; Wilson *et al.*, 1997). The historical impact assessment focuses on two major types of impact.

Chemical impacts on soil development

The chemical processes in soils are complex and can be reflected by a variety of indicators (Condron *et al.*, 1990; Grigal *et al.*, 1991; Johnson, 1992; Wilson *et al.*, 1997). For practical reasons, the assessment of chemical impacts on these processes should be confined to the type and amount of direct input of chemicals such as chalk, fertilizer and pesticides during recent decades, and on the indirect input of atmospheric emissions (average load per hectare). These direct and indirect inputs reflect, to an important extent, the anthropogenically induced changes within the natural soil chemistry.

Physical impacts on soil development

This criterion refers to the time span of physically undisturbed soil development under forest cover. The extent of physical impact is lowest in primary woodland (Prusa, 1985; Smejkal *et al.*, 1997) and in ancient woodland (Ball and Stevens, 1981; Wilson *et al.*, 1997). In this context, deep soil cultivation (affecting the mineral soil) is seen to be the major factor disturbing soil development. In managed forests, the physical impact can be assessed with reference to the last soil cultivation affecting the mineral soil. This assessment aims at distinguishing between soils of high and low physical disturbance at the stand level.

Population dynamics of large herbivores

The numbers of large herbivores, specifically deer, and their natural population dynamic is an important criterion for the naturalness of all processes in woodlands. As it proves difficult to assess the true number of deer per hectare (Andersen, 1953), other indicators, such as the effect of grazing intensity on the natural regeneration of the main tree species, should be chosen to estimate the actual influence of deer on the vegetation dynamics and on its structural development. There are various methods for doing this (Reimoser *et al.*, 1999).

Tree species composition (PNV)

This criterion is a static one, referring to the classical assessment of the PNV (Tüxen, 1956). It reflects the tree species composition and its closeness to the constructed potential natural state of vegetation at a given site. At present the assessment of the PNV is still a matter for intense discussion (see Section 15.2.1) and is put into practice in different ways.

15.5.2 Practical assessment of naturalness

The eight criteria of naturalness reflect a variety of human impacts on a given site. These impacts have to be reconstructed with the greatest possible precision and have to be presented cartographically. This is a question of time and available manpower. The search for adequate information is a single action, that has to be carried out once and then has to be displayed regularly and finally in digitized maps (GIS). The classification of each criterion should simplify the information with regard to the local or regional conditions, e.g. the continuity of forest use, originally displayed in years, can be classified into three grades (high, average, low).

In contrast with some prevailing methods (e.g. Grabherr *et al.*, 1996), the final result of an assessment of naturalness is not a single measure, displayed in different steps, classes or levels, which have weighted and combined the different criteria. It is, rather, a list in which all the criteria are described independently. This provides a maximum of original information that can be combined cartographically according to specific demands and allows one to set up and control specific silvicultural aims (e.g. Westphal, 2001).

It is important to mention that these criteria can be applied to all scales, from the stand to landscape scale. The main goal is to collect information to describe, in a standardized manner, the most important historical influences on woodland dynamics. As there exist infinite combinations of historical and present factors determining the naturalness of a forest, it does not seem useful to evaluate or weigh them in advance, according to a theoretical and subjective procedure. A further evaluation of these criteria should depend on the specific aims of each individual assessment of naturalness.

15.6 The Future

A dynamic understanding of naturalness has fundamental effects on practical forestry. If naturalness is connected to anthropogenically undisturbed woodland processes, then a forestry management aiming to be close to nature has to attempt to reduce human impacts on these key processes. This leads to a longer-term consideration of how to use a forest and what silvicultural methods are applicable to integrate natural dynamics into the present concept of utilization. The assessment of naturalness is therefore not only a tool for evaluation but also for additional management control.

References

Andersen, N.J. (1953) Analysis of Danish roe deer population. *Danish Review of Game Biology* 2, 127–155.

Ammer, U. and Utschick, H. (1982) Methodische Überlegungen für eine Biotopkartierung im Wald. *Forstwissenschaftliches Centralblatt* 101, 60–68.

Aßmann, T. (1998) Bedeutung der Kontinuität von Lebensräumen für den Naturschutz – Untersuchungen an waldbewohnenden Laufkäfern (Coleoptera, Carabidae) mit Beispielen für methodische Ergänzungen zur Langzeitforschung. *Schriftenreihe für Landschaftspflege und Naturschutz* 58, 353–375.

Ball, D.F. and Stevens, P.A. (1981) The role of ancient woodland in conserving undisturbed soils in Britain. *Biological Conservation* 19, 163–176.

Brunet, J. and v. Oheimb, G. (1998) Migration of vascular plants to secondary woodlands in southern Sweden. *Journal of Ecology* 86, 429–438.

Brunet, J., v. Oheimb, G. and Diekmann, M. (2000) Factors influencing vegetation gradients across ancient–recent woodland borderlines in southern Sweden. *Journal of Vegetation Science* 11, 515–524.

Bundesforschungsanstalt für Naturschutz und Landschaftsökologie (1989) Leitlinien des Naturschutzes und der Landschaftspflege in der Bundesrepublik Deutschland. *Natur und Landschaft* 64 (9), 1–16.

Condron, L.M., Frossarde, E., Tiessen, H., Newmann, R.H. and Stewart, J.W.B. (1990) Chemical nature of organic phosphorus in cultivated and uncultivated soils under different environmental conditions. *Journal of Soil Science* 41, 41–50.

Ellenberg, H. Jr (1987) Fülle-Schwund-Schutz: Was will der Naturschutz eigentlich? Über Grenzen des Naturschutzes in Mitteleuropa unter den derzeitigen Rahmenbedingungen. *Verhandlungen der Gesellschaft für Ökologie* 16, 449–459.

Fischer, A. (1995) *Forstliche Vegetationskunde*. Parey, Berlin.

Fritz, Ö. and Larsson, K. (1996) Betydelsen av skoglig kontinuitet för rödliststade lavar. En studie av hallandsk bokskog. *Svensk Botanisk Tidskrift* 90, 241–262.

Grabherr, G., Koch, G., Kirchmeir, H. and Reiter, K. (1996) Wie natürlich ist der Österreichische Wald? *Österreichische Forstzeitung* 1.

Grigal, D.F., McRoberts, R.E. and Ohmann, L.F. (1991) Spatial variation in chemical properties of forest floor and surface mineral soil in the north central United States. *Soil Science* 141, 282–290.

Hanstein, U. and Sturm, K. (1986) Waldbiotopkartierung im Forstamt Sellhorn – Naturschutzgebiet Lüneburger Heide. *Aus dem Walde* 40.

Härdtle, W. (1995) On the theoretical concept of the potential natural vegetation and proposals for an up-to-date modification. *Folia Geobotanica* 30, 263–276.

Hofmann, G. (1997) Mitteleuropäische Wald- und Forstökosystemtypen in Wort und Bild. *AFZ/Der Wald*, special volume.

Hofmann, G. and Jenssen, M. (1999) Quantifizierung der Naturnähe als Planungsgrundlage für praktische Waldumbaumaßnahmen. *AFZ/Der Wald* 54, 575–578.

Jalas, J. (1955) Hemerobe und hemerochore Pflanzenarten. Ein terminologischer Reformversuch. *Acta Societas pro Fauna et Flora Fennoscandia* 72 (11), 1–15.

Jax, K. (1998/99) Natürliche Störungen: ein wichtiges Konzept für Ökologie und Naturschutz? *Zeitschrift für Ökologie und Naturschutz* 7, 241–253.

Jedicke, E. (1998) Raum-Zeit-Dynamik in Ökosystemen und Landschaften. Kenntnisstand der Landschaftsökologie und Formulierung einer Prozeßschutz-Definition. *Naturschutz und Landschaftsplanung* 30 (8/9), 229–236.

Jenssen, M. and Hofmann, G. (1996) Der natürliche Entwicklungszyklus des baltischen Perlgras-Buchenwaldes (*Melico-Fagetum*). Anregung für naturnahes Wirtschaften. *Beiträge für Forstwirtschaft und Landschaftsökologie* 30, 114–124.

Johnson, D.W. (1992) Effects of forest management on soil carbon storage. *Water, Air and Soil Pollution* 64, 83–120.

Karström, M. (1992) Steget Före – en Presentation. *Svensk Botanisk Tidskrift* 86 (3), 103–113.

Knoerzer, D. (1999) Zur Naturverjüngung der Douglasie im Schwarzwald. Inventur und Analyse von Umwelt- und Konkurrenzfaktoren sowie eine naturschutzfachliche Bewertung. *Dissertationes Botanicae* 306.

Koop, H. (1982) Waldverjüngung, Sukzessionsmosaik und kleinstandörtliche Differenzierung infolge spontaner Waldentwicklung. In: Dierschke, H. (ed.) *Struktur und Dynamik von Wäldern*. Berichte der Internationalen Symposien der Internationalen Vereinigung für Vegetationskunde, Vanduz, pp. 235–273.

Koop, H. and Hilgen, P. (1987) Forest dynamics and regeneration mosaic shifts in unexploited beech (*Fagus sylvatica*) stands at Fontainebleau (France). *Forest Ecology and Management* 20, 135–150.

Koperski, M. (1998) Zur Situation epiphytischer Moose in Eichen-Buchenaltbeständen des niedersächsischen Tieflandes. *Forst und Holz* 53(5), 137–139.

Korpel, S. (1995) *Die Urwälder der Westkarpaten*. Fischer Verlag, Stuttgart, Germany.

Kost, G. (1989) Bannwälder als Refugien für gefährdete Pilze. *Natur und Landschaft* 12, 578–582.

Kowarik, I. (1987) Kritische Anmerkungen zum theoretischen Konzept der potentiellen natürlichen Vegetation mit Anregungen zu einer zeitgemäßen Modifikation. *Tuexenia* 7, 53–67.

Mayer, H. (1984) *Waldbau auf soziologisch-ökologischer Grundlage*, 3rd edn. Fischer Verlag, Stuttgart, Germany.

Mayer, H. (ed.) (1987) *Urwaldreste, Naturwaldreservate und schützenswerte Naturwälder in Österreich*. Institut für Waldbau, Vienna.

Michiels, G. (1999) Naturnähe der Waldentwicklungstypen in Baden-Württemberg. *AFZ/Der Wald* 54(16), 866–868.

Mlinsek, D. (1967) Verjüngung und Entwicklung der Dickungen im Tannen-Buchen Urwald Rog (Slowenien). XIV IUFRO – Kongress, Section 23, pp. 436–442.

Neuhäusl, R. (1984) Umweltgemäße natürliche Vegetation, ihre Kartierung und Nutzung für den Umweltschutz. *Preslia* 56, 205–212.

Niedersächsische Landesforstverwaltung (eds) (1994) Langfristige, ökologische Waldbauplanung für die Niedersächsischen Landesforsten. Runderlaß des ML vom 05.05.1994 403/406 F 64210–56.1

Nilsson, C. and Grelsson, G. (1995) The fragility of ecosystems: a review. *Journal of Applied Ecology* 32, 677–692.

Norddeutsche Naturschutzakademie (eds) (1994) Bedeutung historisch alter Wälder für den Naturschutz. *NNA Berichte* 3, 1–159.

Otto, H.J. (1993) Der dynamische Wald. Ökologische Grundlagen des naturnahen Waldbaus. *Forst und Holz* 48, 331–335.

Otto, H.J. (1994) Verminderung der waldbaulichen Intensität und des Schwachholzaufkommens durch naturnahen Waldbau? *Forst und Holz* 49, 387–391.

Otto, H.J (1995) Zielorientierter Waldbau und Schutz sukzessionaler Prozesse. *Forst und Holz* 50, 203–209.

Perpeet, M. (1999) Weniger wäre mehr – waldbauliche Illusion oder Chance? *Forst und Holz* 54(3), 71–74.

Peterken, G.F. (1981) *Woodland Conservation and Management*. Chapman and Hall, London.

Peterken, G.F. (1996) *Natural Woodland: Ecology and Conservation in Northern Temperate Regions*. Cambridge University Press, Cambridge, UK.

Pickett, S.T.A. and White, P.S. (eds) (1985) *The Ecology of Natural Disturbance and Patch Dynamics*. Academic Press, London.

Pickett, S.T.A., Parker, T. and Fiedler, P. (1992) The new paradigm in ecology: implications for conservation biology above the species level. In: Fiedler, P.L. and Jain, S.K. (eds) *Conservation Biology*. Chapman and Hall, New York, pp. 65–88.

Plachter, H. (1995) Functional criteria for the assessment of cultural landscape. In: Droste, B. v., Plachter, H. and Rössler, M. (eds) *Cultural Landscapes of Universal Value*. Fischer Verlag, Jena, Germany, pp. 15–18.

Plachter, H. (1996) Bedeutung und Schutz ökologischer Prozesse. *Verhandlungen der Gesellschaft für Ökologie* 26, 287–303.

Prusa, E. (1985) *Die Böhmischen und Mährischen Urwälder*. Akademia, Prague.

Rackham, O. (1980) *Ancient Woodland: its History, Vegetation and Uses in England*. Arnold, London.

Reif, A. (1999/2000) Das naturschutzfachliche Kriterium der Naturnähe und seine Bedeutung für die Waldwirtschaft. *Zeitschrift für Ökologie und Naturschutz* 8, 239–250.

Reimoser, F., Armstrong, H. and Suchant, R. (1999) Measuring forest damage of ungulates: what should be considered. *Forest Ecology and Management* 120, 47–58.

Remmert, H. (1991) The mosaic-cycle concept of ecosystems – an overview. *Ecological Studies* 85, 1–21.

Rose, F. (1986) Lichenological indicators of age and environmental continuity in woodlands. In: Brown, D.H., Hawksworth, D.L. and Bailey, R.H. (eds) *Lichenology: Progress and Problems. Proceedings of an International Symposium Held at the University of Bristol*. Academic Press, London, pp. 279–307.

Schaal, R. (1994) Waldgeschichtliche Erhebungen im Forstbezirk Münsingen als Beitrag zur Waldbauplanung. *Mitteilungen des Vereins für Forstliche Standortskunde und Forstpflanzenzüchtung* 37, 61–65.

Scherzinger, W. (1996) *Naturschutz im Wald. Qualitätsziele einer dynamischen Waldentwicklung.* Verlag Eugen Ulmer, Stuttgart, Germany.

Schmidt, P.A. (1998) Potentielle natürliche Vegetation als Entwicklungsziel naturnaher Waldwirtschaft? *Forstwissenschaftliches Centralblatt* 117, 193–205.

Schneider, C. and Poschlod, P. (1999) Die Waldvegetation ausgewählter Flächen der Schwäbischen Alb in Abhängigkeit von der Nutzungsgeschichte. *Zeitschrift für Ökologie und Naturschutz* 8, 135–146.

Smejkal, G.M., Bindiu, C. and Visoiu-Smejkal, D. (1997) *Banater Urwälder.* Mirton Verlag, Temeswar, Romania.

Sturm, K. (1993) Prozeßschutz – ein Konzept für naturschutzgerechte Waldwirtschaft. *Zeitschrift für Ökologie und Naturschutz* 2 (3), 181–192.

Sturm, K. (1994) Naturnahe Waldnutzung in Mitteleuropa. Study on behalf of Greenpeace Germany, Hamburg, Germany.

Sukopp, H. (1976) Dynamik und Konstanz in der Flora der Bundesrepublik Deutschland. *Schriftenreihe für Vegetationskunde* 10, 9–26.

Tabaku, V. (1999) Struktur von Buchen-Urwäldern in Albanien im Vergleich mit deutschen Buchen-Naturwaldreservaten und -wirtschaftswäldern. PhD thesis, University of Göttingen, Germany.

Tabaku, V. and Meyer, P. (1999) Lückenmuster albanischer und mitteleuropäischer Buchenwälder unterschiedlicher Nutzungsintensität. *Forstarchiv* 70, 87–97.

Tüxen, R. (1956) Die heutige potentiell natürliche Vegetation als Gegenstand der Vegetationskartierung. *Angewandte Pflanzensoziologie* 13, 5–42.

Usher, M.B. and Erz, W. (1994) *Erfassen und Bewerten im Naturschutz.* Quelle & Meyer, Heidelberg, Germany.

Watkins, C. (1990) *Woodland Management and Conservation,* 1st edn. David & Charles, London.

Westphal, C. (2001) Theoretische Gedanken und beispielhafte Untersuchungen zur Naturnähe von Wäldern im Staatlichen Forstamt Sellhorn (Naturschutzgebiet Lüeneburger Heide). *Berichte des Forschungszentrums Waldökosysteme,* Reihe A, 174, 1–227.

Wilson, B.R., Moffat, A.J. and Nortcliff, S. (1997) The nature of three ancient woodland soils in southern England. *Journal of Biogeography* 24, 633–646.

Integrating Historical Ecology to Restore a Transitional *Pinus palustris* Community

16

G.B. Blank

Department of Forestry, North Carolina State University, North Carolina, USA

Restoration of a Piedmont transitional longleaf pine (*Pinus palustris*) site was undertaken in Wake County, North Carolina. Restoration followed a two-stage examination of land-use history and botanical composition. Factors affecting extant forest composition were identified, and desired characteristics of a restored longleaf stand were posited. Measurement of residual trees after the restoration harvest confirmed gradual, then rapid decline of longleaf recruitment, attributed to infilling by competitor tree species. Radical alteration of stand conditions to liberate longleaf trees resulted in mortality of some residuals but effected changes that would ensure biodiversity associated with longleaf pines in this transitional ecosystem.

16.1 Introduction

In the USA in the past decade or so we have shown great interest in restoring once-prominent but now diminished ecosystems (Egan and Howell, 2001). Whether the interest is whimsical or realistic remains to be seen, but it nevertheless prompts considerable effort to address perceived degraded conditions in North American landscapes. Among ecosystems arousing interest, the once-extensive *Pinus palustris* (longleaf pine) ecosystems of the American South are considered by some observers to be the most endangered (Means, 1996) and therefore deserving of restoration attention (Landers *et al.*, 1995). Given assumed reductions of longleaf pine acreage from 33 or

©CAB International 2004. *Forest Biodiversity: Lessons from History for Conservation*
(eds O. Honnay, K. Verheyen, B. Bossuyt and M. Hermy)

37 million ha (Frost, 1993; Means, 1996) to somewhere between 776,000 to 1.2 million ha today (Landers *et al.*, 1995; Means, 1996), longleaf pine can be considered a poster species for lost forest biodiversity.

The importance of longleaf pine to biodiversity derives from its extensive historic range and its apparent adaptability to diverse sites. The area between Maryland's eastern shore and north-eastern Texas, encompasses longleaf sites with forest co-dominants and understorey species that vary widely. In North Carolina alone, Schafale and Weakley (1990) identify seven distinct longleaf community types; moreover, they note sub-variants. One particular community, the Piedmont transitional longleaf type, receives special scrutiny in this chapter, but is only one of the rare types that have generated interest.

According to Schafale (1994), the Piedmont longleaf pine forest is one of the two rarest longleaf community types in North Carolina. Only nine such sites were included in an inventory by the North Carolina Natural Heritage Program a decade ago (Table 16.1). Along the edge of the Piedmont in North Carolina, longleaf's current significant rarity resulted from clearing for cultivation and pasturage during the 18th and 19th centuries. Fire suppression and foraging by free-ranging swine also contributed to a loss of longleaf dominance in the canopies of sites where it remained. In this transitional zone, few examples of longleaf sites remain today. Hence, discovery in 1987 of a Piedmont tract where longleaf pines persisted within closed-canopy forest prompted attention. The fact that remnant longleaf trees occurred over several hundred hectares drew even more notice because it highlighted past forest conditions generally forgotten by contemporary conservation efforts in the Triangle region (Blank *et al.*, 2002). Longleaf forest of any type only occurs on a few sites in Wake County.

The discovery on the Harris Research Tract (HRT) harkened back to a report published exactly a century before, when Pinchot and Ashe (1897)

Table 16.1. Piedmont longleaf pine forest sites inventoried in North Carolina (Schafale, 1994).

Site	Hectares
Pleasant Grove Church longleaf pine forest	130
Rocky Creek longleaf pine forest	350
Arnett Branch longleaf pine forest	350
Clark's Grove longleaf pine forest	105
Lomax Church longleaf pine forest	140
Goldmine Branch longleaf pine forest	23
Roberdo longleaf pine forest	200
Black Ankle bog	150
West Philadelphia longleaf pine stands	300?

noted abundant longleaf pines in the Apex and Cary vicinity. Given proximity to these towns and the scattered occurrence of longleaf on the Harris Tract, an historical ecology project was begun to examine how current conditions came to be (Bode, 1997). This stage of the project then prompted efforts to characterize the Piedmont transitional longleaf community as it once existed, and to determine the potential for a longleaf community restoration effort (Parker, 1998). Having begun a restoration effort, we are in a position to assess how history affected forest biodiversity and consider how the historic legacy may constrain future biodiversity on this site.

So what happened historically to close the forest canopy, to decrease longleaf pine's dominance, and to eliminate it from portions of the HRT? Does evidence from the plant community tell us about our tract's history? We examined factors of canopy loss and fire suppression as reasons for the loss of longleaf trees. We also devised a list of restoration target species by considering probable and known effects those factors had on species prevalence and dynamics of plant communities. This chapter examines how historical ecology and environmental history played a role in restoring one example of a rare plant community in North Carolina.

16.2 Harris Research Site and Project Methods

The HRT lies in Buckhorn Township of Wake County, North Carolina. Its 513 ha ranges in elevation from 75 to 110 metres above mean sea level. Tributaries of the Cape Fear River, specifically White Oak and Little White Oak Creeks (parts of the Buckhorn Creek watershed) drain the tract. Topography consists of broad, gently sloping ridges divided by a relatively dense pattern of ephemeral and perennial streams. On these uplands, soils that are fine, mixed, semiactive, thermic Typic Hapludults or Aquic Hapludults (Mayodan and Creedmoor, respectively) dominate. They are underlain by, and derived from, material deposited during the Triassic Age. Prior to 2000, existing forest stands ranged from pine of both natural and plantation origin to hardwood of natural origin. Stands of mixed pine and hardwood were most numerous. Longleaf pine had become a minor component in all stands where it occurred.

Aerial photographs of this tract over a 50-year time span, starting in 1938, indicated areas of continuous forest cover and areas of frequent disturbance or total clearance. Since 1988, frequent site visits provided direct observation of changes on the HRT.

A chronology of events and land-use patterns effecting conditions on the HRT prior to purchase by Carolina Power & Light (CP&L) (Table 16.2) was constructed, based on interviews of past residents, documentary research of deeds and land transfer records, and an examination of the agrarian context

Table 16.2.   Historical record pertinent to restoration on the Harris Research Tract (Bode, 1997).

Time period	Activity or tract status
1825	Holleman family begins extracting naval stores in NW corner of the tract
1850	68% of Buckhorn Township classified as unimproved
	82% of Holleman land (out of 281 ha) listed as unimproved
1850–1880	Holleman family grazing cattle, sheep and hogs; decreases land-use intensity
1880	Holleman owns 375 ha; 332 classified as woodland
1895	Patch mosaic of fields and woodland; cessation of fire use
1880–1925	Edwards selectively lumbering in single-tree to ten-acre patches
1890s	New Hill turpentine distillery closes down
1905–1930	Lumbering intensity increased while grazing decreased
1905	Cary Lumbering Company operating on 243 ha
1930–1970	Continuing agriculture with some field abandonment
1938	Aerial photograph series shows open woodlands and scattered fields
1970s	Parcels acquired by Carolina Power & Light (CP&L) for Shearon Harris Facility
1980	CP&L plants open fields with loblolly pines
1988	Aerial photograph series shows all areas forested

and socio-economic characteristics that would effect environmental change in the tract's vicinity (Bode, 1997).

A botanical inventory conducted using the North Carolina Vegetation Sampling protocol (Peet *et al.*, 1998) used 56 tenth-ha (0.1 ha) plots installed across the 513 ha tract, placed to sample all forest types except pine plantations (Parker, 1998). The inventory data, when subjected to cluster analysis, revealed convergence of herbaceous species associated with either Piedmont or Coastal Plain provinces, underscoring the site's transitional nature. Relationships between historical land uses and current plant composition were examined in light of literature review. Ultimately, interpreting existing spatial vegetation patterns and mapping concentrations of remaining longleaf trees (Parker, 1998) focused attention where restoration effort should be directed. One area with the highest concentrations of longleaf pines was identified for the restoration experiment. In the year 2000, a release harvest on 55 ha exposed residual longleaf pines by removing all hardwood and other pines. Two of the original inventory plots in the burned portion of the restoration area were inventoried again during the summer of 2001. On about a third of the restoration site and within the burned area, longleaf pine seedlings were planted in February 2002. All residual tree diameters and heights were measured in May 2002. Cones present and dead trees were also noted; 129 trees were bored to determine numbers of rings at DBH.

16.3 Results

The land-use record constructed for the HRT showed relatively extensive pastoral land use, with intensive agriculture limited to areas most accessible by road, and with varying degrees of forest utilization, from tapping of longleaf trees for turpentine production to selective harvesting (highgrade logging) in old stands (Table 16.2). Although some areas were entirely deforested, some areas appear to have never been entirely cleared. Thus, persistence of longleaf trees on the HRT results from partial cuttings and modest disturbances. The discovery that understorey burning ceased in the late 19th century marks a fairly early departure from that traditional land-use practice, and an important factor for diminished longleaf numbers.

Besides identifying areas where restoration could potentially succeed, the botanical investigation identified a target restoration community (Table 16.3). The target list was based on 12 plots with the highest longleaf pine basal area and comparison with species lists published for other Piedmont longleaf sites (Peet and Allard, 1993; Schafale, 1994).

Measurements of the restoration site's tree component in 2002 revealed 423 trees remaining. Average diameter of the residuals was 36.3 cm, ranging from 72.6 to 11.2 cm ($n = 423$). Average height was 23.3 m, ranging from

Table 16.3. Recommended target species for the Harris Research Tract longleaf restoration site (Parker, 1998).

Dominant canopy	Understorey species	Herbaceous layer	Other species
Acer rubrum	Cornus florida	Gaylussacia	Amelanchier arboreum
Liquidambar	Diospyros	frondosa	Aronia arbutifolia
styraciflua	virginiana	Vitis rotundifolia	Carya alba
Pinus echinata	Nyssa sylvatica		Carya glabra
Pinus palustris	Oxydendron		Chasmanthium laxum
Pinus taeda	arboreum		Chimaphila maculata
Quercus alba	Sassafras albidum		Euonymous americana
Quercus falcata	Vaccinium		Gelssaium sempervirens
Quercus rubrum	corymbosum		Hexastylus arifolia
Quercus stellata			Ilex opaca
			Liriodendron tulipifera
			Lyonia marians
			Prunus serotina
			Quercus phellos
			Rhus copolina
			Smilax glauca
			Smilax rotundifolia
			Symplocus tinctoria
			Vaccinium pallidum

38.43 to 10.4 m ($n = 423$). As of May 2002, 70 trees were dead and 30 had cones in place. Numbers of rings at DBH ranged from 37 to 122, with an average of 82 ($n = 129$). An ice storm in December 2002 that toppled some trees and broke tops from others further reduced the total viable trees by a number undetermined at this point.

Decadal cohorts of the 129 trees bored contain 3% in two classes below 50 annual rings and 12.4% above 100 (Table 16.4). Together, the 70–80 and 80–90 increment classes contain 46.6% of the bored trees. The preliminary study of the post-harvest and -fire herbaceous community on two plots suggests that we will see expected shifts toward earlier succession species. Intermittent drainage ways exposed by the harvest in 2000 exhibit wetland flora observed on some areas of the HRT in 1997 (Parker, 1998), but not in those portions where closed canopy had prevailed.

16.4 Discussion

Decline of longleaf pine as the dominant forest tree of eastern North Carolina and the entire south-east of North America has been widely remarked (Frost, 1990) and lamented (Means, 1996). Thus, extant old growth sites are revered and tended lovingly, their protection virtually guaranteed by their rarity. On the other hand, efforts to restore longleaf pine forests mainly involve planting sites where the species is known to have occurred historically (Landers *et al.*, 1995). Establishing longleaf plantations that prosper on appropriate sites is no longer a problem, especially with intensive competition control. The situation on the HRT lay somewhere between these two poles and prompted a middle course.

16.4.1 Site history

Our aerial photograph series for the HRT suggests that recent conditions, called 'infilled savannah' by Rackham (1998), developed over a span of about

Table 16.4. Decadal cohorts of residual trees.

Cohort	Number	Percentage	Impact decade
≥ 100	16	12.4	1890
< 100 ≥ 90	20	15.5	1900
< 90 ≥ 80	30	23.3	1910
< 80 ≥ 70	30	23.3	1920
< 70 ≥ 60	20	15.5	1930
< 60 ≥ 50	9	7.0	1940
< 50 ≥ 40	2	1.5	1950
< 40 ≥ 30	2	1.5	1960

60 years. Canopy closure resulted from intrusion by species that may have always existed with the longleaf trees but at lower densities, or from intrusion by species that would not have occurred under presettlement conditions. The data in Table 16.4 indicate an abrupt decline in recruitment between 50 and 60 years ago, following five preceding decades of higher recruitment scattered in this sample area. Desire to restore a longleaf community to some semblance of prior diversity and savannah-like conditions demands that we ask about presettlement conditions. Where historic longleaf community composition on the Piedmont is concerned, room for speculation and conjecture based on theory seems inevitable. However, some assumptions about past conditions are safer than others, if based on clues and historic observations bearing the weight of objective scrutiny. In our case, for example, we assume that our transitional site will not resemble wiregrass (*Aristida stricta*) pine savannahs prevalent further south, because our inventory revealed no evidence of wiregrass.

Two traits are fairly consistent across longleaf sites: (i) relatively open canopies and (ii) wide-ranging age classes. Relatively open canopies result from clear dependence of the species on periodic fire regimes to control competition. Platt *et al.* (1988) say 'the absence of stable age-class and size distributions [in longleaf stands] indicates that mortality, growth, and/or recruitment may vary over time' (p. 513). Again, we can note that the air photo series suggests that such characteristics existed as recently as 60 years ago.

16.4.2 Land-use changes

Unlike much of the southern Piedmont, this south-west corner of Wake County was never extensively cleared for intensive row cropping. Tax rolls showed consistently that owners here kept most of their land in an unimproved condition (Trimble, 1974; Bode, 1997). 'Unimproved' meant that the land remained timbered, a prerequiste for longleaf pines to persist. Still, proximity of timber sites to fields cleared for agriculture, and later abandoned, encouraged species able to compete with longleaf. Old-field succession on the North Carolina Piedmont encourages loblolly pine (*P. taeda*), sweetgum (*Liquidambar styraciflua*), and shortleaf pine (*P. echinata*), due to their tolerance of higher light and drier conditions in old fields (Bormann, 1953; Hartnett and Krofta, 1989; De Steven, 1991). Wahlenberg (1946) observes that 'along the northern and eastern edges of the longleaf pine belt, shortleaf pine and loblolly pine often restock former longleaf pine sites. Partly cut stands often have an understory of these two more tolerant species'. We certainly see this potential along margins of the longleaf restoration site. Where loblolly and shortleaf developed undeterred by frequent fires, they 'swamped' longleaf regeneration, and this trend accelerated in future decades (Table 16.4).

The photographic record shows that multiple owners of lands on southern portions of the HRT subjected a higher percentage of their property to intensive agricultural use and thereby eliminated longleaf pine. The central zone of a ridge traversing the HRT north-east to south-west is also nearly completely denuded of longleaf pine, representing historic clearing, probably predating our photographic record.

16.4.3 Woodland management

Disturbance during early decades of the 20th century benefited the longleaf community in places, but subsequent practices favoured other species to the detriment of longleaf pines. Selective logging over several decades on land purchased by Edwards/Cary Lumber Company produced stands with nearly half the basal area of loblolly pine and shortleaf pine, a third that of sweetgum, and approximately six times that of southern red oak, compared to areas where the canopy was entirely removed.

Where woodland persisted on the HRT, practices of harvesting pine resin (turpentining), hog foraging and fire suppression also negatively impacted longleaf pine. Results of these practices differed in degree of degradation. Between 1825 and 1880 pine resin (pitch) was collected for a turpentine distillery, operating in New Hill until the 1890s. Collecting pine resin for turpentine did not necessarily remove all longleaf pines, but the average longleaf tree would only produce for 10–15 years. Then depleted trees were typically harvested for destructive distillation, removing mature trees, hence seed sources. A common practice to protect tapped trees from accidental fire was to maintain vegetation-free zones around the producing trees (Croker, 1979). Such zones would have reduced numbers of longleaf pine seedlings. Moreover, devastation of young longleaf (seedlings and saplings) by foraging hogs has been well documented. We know that hogs were raised on the HRT, although we cannot ascertain the exact boundaries of their effects. Traces of hog fencing throughout the woodland and broad portions of landscape lacking longleaf trees suggest either old fields already reforested before the earliest aerial photographs or areas where hogs may have done their damage.

However, on the whole, areas where longleaf pines persist did not appear to suffer the heavy naval stores activities or degradation by hog grazing (Parker, 1998). Absence of these activities, and proximity to viable seed of mature longleaf pines, allowed longleaf pine to exploit the same openings as shortleaf pine and loblolly pine. Openings are vital to longleaf pine regeneration because, as Wahlenberg (1946) noted: 'root competition, rather than shade, accounts for the absence of seedlings under the open crowns of trees'. Openings would have to be large enough to allow seedlings to emerge and prosper beyond roots of parent trees. Exploitation of openings was still occurring on the HRT in the 1990s, as exhibited by natural regeneration found within some inventory plots, gaps along woods road margins, old fire lines, and tree falls.

Over decades, longleaf has clearly lost ground to its pine competitors; the best evidence for this may come from a hardwood species. Portions of HRT with high basal areas of *Quercus falcata* confirm a history of fire suppression as opposed to canopy loss (Parker, 1998). In a frequent fire regime, *Q. falcata* populations are suppressed, but populations dramatically increase when fire frequency decreases. Frost (1998) asserts that 'fire suppression for as little as a decade results in substantial increases in the size and abundance of hardwood stems within longleaf forests'. We conclude that the greatest negative impact on longleaf dominance in continuously-forested areas resulted from fire suppression.

Oldest cohorts far surpass youngest cohorts in numbers of representatives. During more recent decades, as stand density increased with infilling by other species, longleaf regeneration diminished and seedlings that might have become established remained suppressed by increasing competition from trees around them.

16.4.4 Restoration path

On the HRT, preservation of the *status quo* would eventually lead to total loss of the longleaf component in this forest. Totally harvesting the existing stand of timber and replanting longleaf seedlings would have eliminated legacy trees. Desire to allow extant trees to regenerate the site, yet the need to capture the site with a viable longleaf population, led to the current course of action. We have some old trees and we have young seedlings. At issue still is what will happen to the rest of a diverse plant community surrounding these trees. We wonder whether, in the aftermath of the decades-long movement away from natural, the stand will approach what observers posit as characteristic of 'natural' longleaf communities.

Lacking a clear reference site, we are working with a target template. The limited number of remaining Piedmont longleaf sites means that the conservation community does not know much about species composition of North Carolina's original Piedmont transitional longleaf community (Shafale, 1994). Degraded mature longleaf sites available for examination as reference sites offer limited evidence due to intervening land-use practices (Murray, 1995; Cohen, 1998). Estimates made from a combination of historical documentation, neighbouring identified forest types, and current species found on the Harris Tract are viewed in the frame of forest succession the tract has undergone. Species lists from other Piedmont longleaf sites (Peet and Allard, 1993; Schafale, 1994) prompted Parker to make comparisons (Table 16.5) and to construct a target list for this site (Table 16.3).

Subsequently introducing prescribed burning is intended to foster longleaf regeneration and favour herbaceous species, possibly inhabiting the site prior to original disturbance events and the later cessation of intermittent fires. A

Table 16.5. Species found at low densities on the Harris Research Tract and also noted by other studies.

Peet and Allard (1993)	Schafale (1994)
Aronia abutifolia	*Pteridium aquilinum*
Asplenium platyneuron	*Quercus marilandica*
Eupatorium rotundifolium	*Silphium compositum*
Euphorbia corollata	*Solidago odora*
Hypericum hypericoides	*Tephrosia virginiana*
Myrica cerifera	*Vaccinium arboreum*
Pityopsis graminifolia	*Vaccinium tenellum*
Quercus nigra	

19-ha longleaf plantation nearby, after its third growing season, demonstrates viability of the species under relatively intensive management. Our multiple-age restored stand will, we hope, provide a laboratory for examining dynamics of the mixed forest composition, relatively open canopy structure, and herbaceous biodiversity, once prevalent on much larger portions of the Piedmont than exist today, or will likely ever exist again.

16.4.5 Conclusions

The historic legacy on the Harris Research Tract offers a chance to reaffirm the important complex of community interactions occurring within the realm of longleaf pine ecosystems. Further studies of the developing plant community will proceed, and the fate of residual trees will be monitored. But what we have seen thus far suggests a need for more far-reaching inventory work in the historically transitional longleaf regions of the south-eastern USA. On sites where remnants of the longleaf pine community occur, though perhaps not in significant numbers to stand out from surrounding forest, the species has been missed. Opportunities for restoration are being lost through general neglect of conditions and practices that would revive significant portions of the once dominant transitional longleaf plant community. To some extent, some of these sites could be renovated even as forest tracts fragment from suburbanization. But if we do not identify where they are, we have no hope of restoring them. Meanwhile, available data skew toward a much more pessimistic view of our environment than is necessary if we are, communally and collectively, serious about achieving measurable degrees of sustainability.

References

Blank, G.B., Parker, D.S. and Bode, S.M. (2002) Multiple benefits of large, undeveloped tracts in urban landscapes. *Journal of Forestry* 100 (3), 27–32.

Bode, S.M. (1997) Land use and environmental history of the Shearon Harris Tract. Masters thesis, North Carolina State University, Raleigh, North Carolina.

Bormann, F.H. (1953) Factors determining the role of loblolly pine and sweetgum in early old-field sucession in the Piedmont of North Carolina. *Ecological Monographs* 23, 229–358.

Cohen, S. (1998) Restoration potential of disturbed longleaf pine understorey plants. Masters thesis, North Carolina State University, Raleigh, North Carolina.

Croker, T.C. (1979) Longleaf: the longleaf pine story. *Journal of Forest History* (January), 32–43.

DeSteven, D. (1991) Experiments on mechanisms of tree establishment in old-field succession: seedling emergence. *Ecology* 72, 1066–1075.

Egan, D. and Howell, E.A. (2001) *The Historical Ecology Handbook: a Restorationist's Guide to Reference Ecosystems.* Island Press, Washington, DC.

Frost, C.C. (1990) Natural diversity and status of longleaf pine communities. In: *Forestry in the 1990s: a Changing Environment.* Regional Technical Conference and 69th Annual Business Meeting of the Appalachian Society of American Foresters, Pinehurst, North Carolina.

Frost, C.C. (1993) Four centuries of changing landscape patterns in the longleaf pine ecosystem. In: *The Longleaf Pine Ecosystem: Ecology Restoration and Management.* Proceedings, 18th Tall Timbers Fire Ecology Conference. Tall Timbers Research Station, Tallahassee, Florida.

Frost, C.C. (1998) Presettlement fire frequency regimes of the United States: a first approximation. In: Pruden, T.L. and Brennan, L.A. (eds) *Fire in Ecosystem Management: Shifting the Paradigm from Suppression to Prescription.* Tall Timbers Ecology Conference Proceedings, 20th Tall Timbers Research Station, Tallahassee, Florida, pp. 70–81.

Hartnett, D.C. and Krofta, D.M. (1989) Fifty-five years of post-fire succession in a southern mixed-hardwood forest. *Bulletin of Torrey Botanical Club* 116, 107–113.

Landers, J.L., Van Lear, D.H. and Boyer, W.D. (1995) The longleaf pine forests of the southeast: requiem or renaissance? *Journal of Forestry* 93, 39–44.

Means, D.B. (1996) Longleaf pine forest, going, going.... In: Davis, M.B. (ed.) *Eastern Old-growth Forests: Prospects for Rediscovery and Recovery.* Island Press, Washington, DC, pp. 210–229.

Murray, A.N. (1995) Description and restoration of a relict longleaf pine flatwood in Halifax County, North Carolina. Masters thesis, North Carolina State University, Raleigh, North Carolina.

Parker, D.S. (1998) Using botanical analysis to shape a longleaf restoration project. Masters thesis, North Carolina State University, Raleigh, North Carolina.

Peet, R.K. and Allard, D.J. (1993) Longleaf pine vegetation of the Southern Atlantic and Eastern Gulf Coast regions: a preliminary classification. In: *The Longleaf*

Ecosystem: Ecology, Restoration and Management. Proceedings, 18th Tall Timbers Fire Ecology Conference. Tall Timbers Research Station, Tallahassee, Florida.

Peet, R.K., Wentworth, T.R., Duncan, R. and White, P.S. (1998) A flexible, multipurpose method for recording vegetation composition and structure. *Castanea* 63, 262–274.

Pinchot, G. and Ashe, W.W. (1897) *Timber Trees and Forests of North Carolina.* Geological Survey Bulletin 6, M.I. and J.C. Stewart Public Printers, Winston, North Carolina.

Platt, W.J., Evans, G.W. and Rathbun, S.L. (1988) The population dynamics of a long-lived conifer (*Pinus palustris*). *American Naturalist* 13, 491–525.

Rackham, O. (1998) Savanna in Europe. In: Kirby, K.J. and Watkins, C. (eds) *The Ecological History of European Forests.* CAB International, Wallingford, UK, pp. 1–24.

Schafale, M.P. (1994) *Inventory of Longleaf Pine Natural Communities in North Carolina.* North Carolina Natural Heritage Program, Division of Parks and Recreation, Department of Environment, Health, and Natural Resources, Raleigh, North Carolina.

Schafale, M. and Weakley, A.S. (1990) *Classification of the Natural Communities of North Carolina: Third Approximation.* North Carolina Natural Heritage Program, Division of Parks and Recreation, Department of Environment, Health, and Natural Resources, Raleigh, North Carolina.

Trimble, S.W. (1974) *Man-induced Soil Erosion on the Southern Piedmont 1700–1970.* Soil Conservation Society of America.

Wahlenberg, W.G. (1946) *Longleaf Pine, its Use, Ecology, Regeneration, Protection, Growth, and Management.* Charles Lathrop Pack Forestry Foundation and USDA Forest Service, Washington, DC.

Is the US Concept of 'Old Growth' Relevant to the Cultural Landscapes of Europe? A UK Perspective

17

K.N.A. Alexander and J.E. Butler

Ancient Tree Forum, c/o Woodland Trust, Autumn Park, Grantham, Lincolnshire, UK

The term 'old growth' was originally initiated in the USA, where it has been applied to relatively undisturbed natural forest ecosystems characterized by large, old trees. It might, therefore, be considered that Europe's cultural landscapes are automatically excluded from the definition. However, if old growth is defined in terms of the richness of its native species associations, then many European sites could be considered to be old growth, albeit used by people over extended time periods. This chapter will discuss the species approach to defining old growth, with emphasis on the saproxylic, epiphytic and mycorrhizal communities, with specific site examples from the UK and elsewhere in Europe.

17.1 Introduction

The term 'old growth' originated in the western USA as a means of identifying forest areas with large old trees (IUFRO, 1997). Over the years there have been many attempts to define what is actually meant by the term (Lund, 2001). These usually rely on structural characteristics, but identified areas are always implicitly species-rich, with high-profile species present, such as the northern spotted owl or the fungus *Bridgeoporus nobilissimus* (Castellano *et al.*, 1999).

'Old growth' has increasingly come into use in the UK and elsewhere in Europe. But is it really relevant to our cultural landscapes? In 2002 English Nature (EN) (the UK Government agency responsible for nature conservation

©CAB International 2004. *Forest Biodiversity: Lessons from History for Conservation*
(eds O. Honnay, K. Verheyen, B. Bossuyt and M. Hermy)

in England) awarded a contract to the Ancient Tree Forum (ATF) (an independent charity) to explore the issues behind the term and examine whether or not the term had any real validity for the UK in particular, but also to consider the international European context. Specific objectives were to develop a working definition of old growth for the UK, and a practical method for developing an inventory of sites. A final report has been published in the *English Nature Research Reports* (Alexander *et al.*, 2003).

17.2 Convention on Biological Diversity Definition

The carefully worded international definition provided on the Convention on Biological Diversity (CBD) website (www.biodiv.org/convention/articles.asp) appears to be the most widely accepted modern definition of what is meant by old growth:

> Old growth forest stands are stands in primary or secondary forests that have developed the structures and species – normally associated with old primary forest – of that type that have sufficiently accumulated to act as a forest ecosystem distinct from any younger age class.

This has been identified by the EN project as an adequate starting point for consideration of the UK and European context.

17.3 UK Situation

In the UK and the rest of Europe there is arguably no virgin forest left. Peterken (1996) defines virgin forest in the following terms: 'natural woodland which has never been significantly influenced by people'. The landscapes of the entire continent are most appropriately regarded as cultural landscapes. People have been using and exploiting the natural vegetation cover for thousands of years (e.g. Rackham, 1986, 1990; Birks *et al.*, 1988; Salbitano, 1988; Kirby and Watkins, 1998; Butler *et al.*, 2002). There is virtually a complete spectrum of modification of the natural forest, from – at one extreme – completely treeless lands, through to still well-wooded country populated with species characteristic of the former primary woodland cover of the land.

17.4 How Should We Interpret the CBD Definition?

Breaking down the CBD definition into its component parts helps to clarify the intent. 'Primary or secondary' makes it clear that the definition does not confine itself to virgin or little-modified forests and allows the inclusion of today's European cultural landscapes. Peterken (1996) defines primary woodland as 'woodland and wood pasture which has existed continuously since before the original forests were fragmented' and secondary woodland as 'woods and wood pastures which have originated on un-wooded ground since the original forests were fragmented'.

'Structures and species associated with old primary forest' leads us into difficulties. Vera (2000) has recently presented a strong case that we have been very wrong in our interpretation of the evidence for the structure of old primary forest, and makes the case for a new understanding of a more open structure. But whatever the structure actually was, it can probably be recreated in relatively short time scales, of perhaps only a few hundred years. Recent research, for instance, has shown that quite considerable gains in biodiversity can be made by increasing volumes of dead wood in plantations – although the benefits are most obvious in common and widespread species (Humphrey *et al.*, 2002). Some of the rarer and more exacting species have also been found following this treatment, and this might be seen as an encouraging result, but there is still no real evidence that relict old forest species are capable of re-colonizing in any numbers. It is extremely difficult to identify how these rarer species were able to reach the study sites. Humphrey *et al.* (2002) also comment that these plantation reserves can provide a *semblance* (our italics) of old growth habitat conditions, including large trees, structural diversity and deadwood.

The requirement of the CBD definition for species associated with primary forest is much more exacting. Species richness cannot so readily be restored, and there is a strong case to be made that the species content of a particular area is the only proof of old growth conditions.

Species which have been proposed as particularly characteristic of relict areas of old primary forest in the UK include three major ecological groupings (Butler *et al.*, 2001):

1. Epiphytes, especially lichens, but also bryophytes and the invertebrates associated with vegetated tree bark.
2. Saproxylic fungi and invertebrates.
3. Mycorrhizal fungi and their associated invertebrate fauna.

Vascular plants also include many species which are characteristic of old forest (for example, see Peterken, 1996). Good examples are the wild service tree (*Sorbus torminalis*), small-leaved lime (*Tilia cordata*) and herb Paris (*Paris quadrifolia*). However, the ground vegetation species tend to occur with lesser prominence than in enclosed and ungrazed closed canopy woodlands. Birdlife (especially woodpeckers) is also an important feature of old forest.

The three groupings of organisms – epiphytic, saproxylic and mycorrhizal communities – are well known for their strong association with ancient open-grown trees, trees managed historically as pollards, areas with concentrations of ancient trees, and areas with a long-documented history of old trees. Species richness of these groups of organisms is associated with ancient wood pastures and parklands (Harding and Rose, 1986; Watling and Ward, 2003) – areas where traditional extensive grazing and wood-cutting has continued to mimic the natural processes of primary forest.

The inescapable conclusion is that concentrations of ancient trees should be recognized as the most practical structural indicator that we have of areas that have exceptionally species-rich biological communities, and continuity with old primary woodland. Ancient trees should be recognized as icons of old growth communities.

17.5 How Long Does it Take to Achieve Old Growth?

Many definitions of old growth have sought to identify criteria related to tree age, usually focusing somewhere in the range of 120–250 years, with 200 years being the most commonly quoted (Lund, 2001). Peterken (1996) suggested a general European application of the term to stands of trees that show 200 years of undisturbed growth, but later reduced this to 150 years (Peterken, 2000). But, of course, different tree species have different life spans and so arbitrary time limits are inappropriate (Butler *et al.*, 2001). Oaks usually only reach biological (as opposed to commercial) maturity at 300 years of age, beech is entering its post-mature stage at 200 years, whereas birch, in most circumstances in the lowlands, would be considered ancient at 100 years.

A more useful approach would be to define old growth in terms of a set time after commercial maturity is exceeded for the species concerned. Stands exceeding the regular cycle by 20 years have been suggested for the boreal forests of Finland (Lund, 2001). Thompson and Angelstam (1999) summarize work in boreal forests as indicating that particularly high diversities of old growth-dependent species occur in stands 50–100 years older than the age at which trees are normally harvested (usually 80–120 years).

And yet, in order to successfully conserve the whole suite of old forest species – including species which require ancient trees – it is essential that ancient trees are well represented in the areas under conservation management. It follows, therefore, that old growth reserves need to be identified on the basis of concentrations of ancient individuals of the different tree species represented.

Also, if the survival of species characteristic of primary forest is to be assured, then the generation span of trees needs to be considered – overlapping generations of trees are essential for transfer of the associated species, and this is particularly exacting with species that depend on ancient trees. Butler *et al.* (2001) consider that continuity, by which they mean 'a continuous presence of old trees dating back into the past' or 'dating back to original forest cover', should be an additional criterion for categorizing 'old growth' woodland. They cite evidence that biodiversity values (lichen and Coleoptera species-richness) are reduced in isolated sites where old tree cover has been interrupted (e.g. Rose, 1992, 1993; Alexander, 1998). The latter, in particular, demonstrates the relationship between continuity and species richness in saproxylic Coleoptera (see below).

The proposed age-related criteria are also much too short for many tree species. Under growing conditions in much of the UK the native pedunculate oak (*Quercus robur*) normally only begins to be colonized by heartwood decay fungi after 150–200 years, and so tree-hollow species can only colonize much later, perhaps after 250–300 years. The UK's rarest cobweb beetle, *Trinodes hirtus*, requires even older trees, as it needs the development of extensive cavities within the outer tree bark which have been colonized by large populations of web-forming spiders, particularly *Amaurobius* spp. (Alexander, 2002). Basically, 200–250-year-old oaks have only just started to be colonized by some of the key species of old primary forest.

Within the definition provided by the CBD there is an acceptance that secondary woodlands can develop the structures of old primary woodland. However, there is great emphasis in the definition on the development of an ecosystem with structure and species, distinct from any younger age class. Structure may develop over time, and natural processes may become established, but if the secondary forest is isolated from species-rich old primary forest, it is unlikely to develop the species richness of old primary forest.

Old growth in the structural context is defining a stage in a process, and this is always going to be a snapshot in a stand's history. Ancient trees are ultimately going to be temporary individuals in any wooded landscape (although on a scale of centuries, not decades), whether it is in the coniferous shade-tolerant forests of the boreal or Pacific Northwest or the open ancient wood pastures of temperate oaks. In one sense this is the strength of using ancient trees as a structural indicator of old growth – the stands are relatively easily defined, both conceptually and on the ground. Should it matter, however, if the regenerating cohort is either a separate individual or distinct patch, or, indeed, not yet present but awaiting a temporary break in grazing pressure to get established? The important criterion is that the species associated with the old trees can survive in the ecosystem, which means that we need continuity of habitat reaching into the past. The continuous presence of old trees is less easy to define, but this is where the indicator species approach provides a way forward.

17.6 Continuity and Indicator Species

It is relatively straightforward to identify ancient trees. However, demonstrating continuity of ancient trees in an ecosystem – reaching back into the past – is exacting and requires assessment by specialist ecologists. In primary old forest, continuity is implicit and does not need to be separated out. In our modified and fragmented cultural landscape, however, continuity of habitat is believed to be an important factor in the conservation of certain relatively immobile species and/or rare communities.

The presence of species of old primary forest will always indicate that there is sufficient ecosystem structure and processes, i.e. habitat, present. Even where structure and process is very limited, e.g. just old trees standing in very modified environments, the degree of presence or absence of these particular species and/or communities has been successfully used to assess the level of continuity.

These special species include: saproxylic fungi such as *Piptoporus quercinus*, *Hericium erinaceum* and *Podoscypha multizonata*; mycorrhizal fungi such as *Boletus satanas*, *Cortinarius violaceus* and *Amanita ceciliae*; epiphytic lichens such as *Lobaria virens*, *Arthonia vinosa* and *Thelotrema lepadinum*; saproxylic beetles such as *Osmoderma eremita*, *Limoniscus violaceus* and *Cerambyx cerdo*; saproxylic hoverflies such as *Caliprobola speciosa*, *Callicera spinolae* and *Myolepta potens*, and the saproxylic cranefly *Ctenophora flaveolata*.

Just how many of the recognized, or acknowledged, old forest species are needed to assess a particular area as having old growth will require further debate. The various indices that have been developed (see later) do provide threshold levels, above which a particular site is said to have regional, national or even international significance. Threshold indices do seem to be a practical way forward. Sites with the essential structures of old primary forest, i.e. ancient trees, but which have not yet been studied for their species composition, can be regarded as 'provisionally' old growth until the necessary research is carried out.

The species-richness approach developed in Sweden – Signalarter (Nitare, 2002) – neatly side-steps the issue of continuity. Selections of species are used as indicators for assessing the nature conservation value of woodland sites. However, this straightforward approach has not been accepted in the UK, where we are locked into a long-running debate over whether or not there is reliable evidence that particular species can be used – either singly or collectively – as indicators of continuity.

The evidence for continuity is, of course, circumstantial, given our present inadequate knowledge of the species composition of the primary forest of the UK, or elsewhere in Europe. However, the suites of species that have been proposed as indicators are validated by a wide variety of circumstantial evidence and, collectively, this is considered to be sufficiently persuasive by many experts:

- research on site history;
- Holocene sub-fossil record; and
- species mobility.

17.7 Research on Site History

Rose's investigations (1974, 1976, 1988) of the relationship between wood-lands of known age and lichen species-richness involved a large number of sites across Britain and France. In the main, indexes are higher in wood–pasture scenarios where considerable numbers of ancient trees have survived. This work has established the break in continuity independently of the species in a significant number of sites. More detailed research on the epiphytic lichen flora of chronosequences (stands of known age) back to 600 years old (Sanderson, 1996, 1998) has shown that the lichen indices do reflect continuity in the New Forest. However, this continuity is at a site level, not stand level, and continuity is destroyed by fragmentation of more than a kilometre or so over this time scale.

Alexander (1988, 1998) has analysed similar evidence for saproxylic Coleoptera across Britain. Figure 17.1 (taken from Alexander, 1998) demonstrates the relationship between species-richness and site history very clearly. The richest sites for saproxylic Coleoptera are all well documented as having a long and unbroken history of suitable habitat. Sites which, on the inadequate historical evidence available at the time, appeared to go against this trend have been eroding away as new historical or archaeological data becomes available. The area of today's Moccas Park (Herefordshire) has recently been shown to have held ancient wood pasture back into the medieval period (Harding and Wall, 2000) while the area of today's Dunham Massey Park (Manchester) is now known to have been a medieval deer park serving the medieval castle of Dunham (National Trust archaeological survey and archive data). The area of today's Knole Park (Kent) was open wood pasture in the medieval period.

17.8 Holocene Sub-fossil Record

Sub-fossil studies provide a reference point to some extent, although informa-tion from this source is still very incomplete. However, knowledge of the number of extinctions of saproxylic Coleoptera following the break-up of the natural forest cover of the UK continues to grow (see, for example, Buckland and Dinnin, 1993). The impact of the increasing fragmentation and isolation of individual surviving populations is also being increasingly demonstrated; Buckland (2002), for example, provides examples of species currently showing an extreme relict old forest distribution which were more widespread in the Neolithic period. Examples include the weevil *Dryophthorus corticalis* and the longhorn beetle *Cerambyx cerdo* – associated with old decaying oak, and

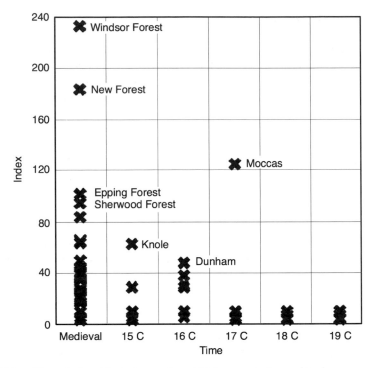

Fig. 17.1. The relationship between forest history and site quality for saproxylic Coleoptera (from Alexander, 1998). The Index refers to the Index of Ecological Continuity (Alexander, 1988) which is calculated from site lists using species selected and graded by Harding and Rose (1986). The named sites all lie within lowland England: Windsor Forest and Great Park, Berkshire; New Forest, Hampshire; Epping Forest, Essex; Sherwood Forest, Nottinghamshire; Knole Park, Kent; Dunham Park, Manchester; and Moccas Park, Herefordshire.

Rhysodes sulcatus and *Isorhipis melasoides* with decaying beech (*Fagus sylvatica*). *Dryophthorus* once extended widely across England but is now confined to Windsor Forest and Great Park. *Rhysodes* and *Isorhipis* were once similarly widespread in England but both are now extinct there.

Research on fossil insect assemblages has provided good evidence that key beetle indicator species are indeed species of primary forest before it was disturbed by human activity. Buckland and Dinnin (1993) have concluded that the remaining refuges for old-forest insects are the ancient wood pastures, and that these species are hanging on in areas of parkland – 'a managed habitat which is perhaps the oldest in Britain'. Most of these ancient wood pastures and parklands have been poorly documented in the past but are now regarded as a priority habitat under the UK Biodiversity Action Plan, and progress is now being made.

17.9 Species Mobility

Species mobility in epiphyte, saproxylic and mycorrhizal communities is notoriously difficult to study. A strong correlation between relict old forest species and relatively low mobility values has long been suspected and there is much circumstantial evidence. However, quantitative studies are rare, but some work has recently been carried out on lichens and beetles (Sanderson, 1996, 1998; Alexander, 2003; Brunet, Chapter 10, this volume).

Old forest lichens and invertebrates have been demonstrated colonizing suitable trees *within* old wood pastures and parklands, and extending short distances outside of the fragment of their habitat in which they live; evidence for movement *between* sites is extremely rare, and evidence for colonization of *new* sites is non-existent. Sanderson's (1996, 1998) work on lichens, for example, has shown that mobility tends to be at a site level. Alexander (2003) has recently shown that one saproxylic beetle species, *Gastrallus immarginatus*, has not been able to cross a distance of just 500 m between stands of suitable *Acer campestre* host trees in about 400 years. Brunet (Chapter 10, this volume) presents further evidence from Sweden that epiphytic lichens and saproxylic Coleoptera demonstrate severe dispersal limitation at distances further than about 200 m.

Figure 17.1 provides evidence from a different route. There are two ways of interpreting this evidence for mobility in saproxylic Coleoptera. One option assumes that sites documented back to the medieval period have virtually guaranteed continuity right back to the primary forest cover of the area and therefore these species have not been able to colonize new sites over that extended time period. A more cynical interpretation would be that these species have not been able to colonize new sites in the 500 or so years between the medieval period and today. Either way, the time period involved for colonization lies in the range of 500–1000s of years, which is considerably longer than the time period for the development of old growth structures.

17.10 Ascertaining Continuity

Ascertaining continuity on a wide range of sites requires specialist survey backed by historical research (Selva, 1994). The existing survey methods use the richness of the more immobile lichen, invertebrate and fungal communities – with their highly specialized requirements – to create a reference index of ecological continuity such as those devised by Rose (1976, 1993) and Alexander (1988; updated in Harding and Alexander, 1994). The species assemblages associated with epiphytic and saproxylic habitats have been analysed and the individual species categorized according to the extent of their association with the most species-rich localities, i.e. those localities known to support the most complete suite of species for the habitats represented. A selection of the species associated with the most species-rich sites has then been identified for use as indicator species, and their cumulative presence used to calculate indices of ecological continuity. These appear to be adaptable across western European deciduous woodland providing supporting evidence for a link to the primary forest. Speight (1989) provides a list of about 200 species of saproxylic insects applicable across Europe.

Regional variation in lists of indicator species needs to be recognized. Species may behave in different ways under different climatic conditions, and different regions will support different suites of species. There are many examples. Amongst saproxylic beetles: *Bitoma crenata* is a stronger indicator of continuity in the west than in the south-east of Britain; *Agrilus biguttatus* is still confined to Sherwood Forest in the English midlands but has recently become highly mobile across the south-east of the country, most probably in response to climate change; *Orchesia undulata* has a very restricted distribution in Germany and the Czech Republic but is widespread over much of Britain. The epiphytic lichen *Lobaria pulmonaria* is an excellent indicator species in lowland Britain (Rose, 1976) and southern Sweden (Nitare, 2000) but less reliable in the Atlantic zone of the far west. Hazel scrub in the north-west of Scotland supports a different suite of lichen species than the southern beech woods and this variation is reflected in the choice and range of indicator species (Hodgetts, 1992).

17.11 Need for Conservation Action

In many ways, discussions about ecological continuity and old growth definitions are rather academic. Everything that we have learnt about sites with ancient trees demonstrates that these sites are amongst the most:

* species-rich of all European woodlands, especially for saproxylic, epiphytic and mycorrhizal communities;
* important;
* vulnerable; and
* neglected by the conservation movement.

We urgently need action to protect sites with ancient trees at a European level. The first priority must be to identify species-rich sites for old growth and protect them. Site quality assessment should be based on:

* concentrations of ancient trees, as our most reliable structural indicator; or
* knowledge of epiphytes, saproxylics and mycorrhizals.

Action is also urgent at the policy level:

* EU level priorities for old growth;
* protection of existing areas; and
* buffering and extending existing reserves.

Speight (1989) made these same recommendations over 10 years ago – why has action been so poor? There are many possible reasons. A particular problem has been the lack of clear definition of terms, which has resulted in poor understanding of the terminology and awareness of the issues amongst ecologists, let alone among the people working at a high policy level with the politicians. Old growth has long been known to be of importance by specialist biologists, people who publish in their own narrow discipline journals and fail to promote the wider multidisciplinary aspects amongst a more general audience. A key problem has been the failure of ecologists and nature conservationists to see beyond the conventional view of narrowly defined semi-natural habitats such as 'woodland' and recognize the importance of habitat mosaics, of trees growing – singly or in groups – within a matrix of grassland, heathland and other open habitats. The way forward is to think in terms of landscapes of trees and woodlands – treescapes. Some progress has been made in the UK (see, for example, Alexander, 2000) but this needs to be consolidated and extended. Greater vision is needed amongst the policy makers (Tubbs, 1997).

Where there are sites that retain unusually large numbers of species associated with intact ecosystems, the special biodiversity value of these sites needs to be internationally recognized, greater emphasis given to their conservation and their value promoted to the public. There is a need for a special term or expression to emphasize the value of these ecosystems and in promoting the values to wider audiences. 'Old growth' provides this expression.

17.12 Conclusions

Europe's old growth is in its cultural landscapes, particularly its wood pastures and historic parklands with ancient trees. Ancient trees conserve many of the relict species of European old primary forest. This interpretation is consistent with the Convention on Biological Diversity definition of old growth. Ancient trees should be recognized as the focus for conservation action on old growth in the UK and across Europe as a whole.

In our fragmented landscape, every area of old growth is precious, and merits identification and special protection measures. The conservation value of any stand is as much a factor of the wider habitat in which it is found as of its intrinsic structure. Thus protection initiatives must be at landscape level.

References

Alexander, K.N.A. (1988) The development of an index of ecological continuity for deadwood associated beetles. *Antenna* 12, 69–70.
Alexander, K.N.A. (1998) The links between forest history and biodiversity: the invertebrate fauna of ancient pasture–woodlands in Britain and its conservation.

In: Kirby, K.J. and Watkins, C. (eds) *The Ecological History of European Forests*. CAB International, Wallingford, UK, pp. 73–80.

Alexander, K.N.A. (2000) Action to conserve ancient trees and their associated wood-decay communities at landscape level. In: *Workshop on Ecological Corridors for Invertebrates: Strategies of Dispersal and Recolonisation in Today's Agricultural and Forestry Landscapes*. Neuchâtel (Switzerland), 10–12 May 2000. Environmental encounters No. 45, Council of Europe Publishing, Strasbourg, France, pp. 61–65.

Alexander, K.N.A. (2002) The invertebrates of living and decaying timber in Britain and Ireland – a provisional annotated checklist. *English Nature Research Report* No. 467, Peterborough, UK, pp. 1–142.

Alexander, K.N.A. (2003) *Gastrallus immarginatus* (Müller) (Anobiidae) at Hatfield Forest, Essex. *The Coleopterist* 12.

Alexander, K., Smith, M., Stiven, R. and Sanderson, N. (2003) Defining 'old growth' in the UK context. *English Nature Research Report* No. 494, Peterborough, UK, pp. 1–54.

Birks, H.H., Birks, H.J.B., Kaland, P.E. and Moe, D. (eds) (1988) *The Cultural Landscape. Past, Present and Future*. Cambridge University Press, Cambridge, UK.

Buckland, P.C. (2002) Conservation and the Holocene record – an invertebrate view from Yorkshire. In: *Recording and Monitoring Yorkshire's Natural Environment*. Supplement to Bulletin Yorkshire Naturalists' Union No. 37.

Buckland, P.C. and Dinnin, M.H. (1993) Holocene woodlands, the fossil insect evidence. In: Kirby, K.J. and Drake, C.M. (eds) *Dead Wood Matters: the Ecology and Conservation of Saproxylic Invertebrates in Britain*. English Nature Science No. 7, Peterborough, UK, pp. 6–20.

Butler, J., Alexander, K. and Green, E. (2002) Decaying wood – an overview of its status and ecology in the United Kingdom with some remarks on continental Europe. In: Laudenslayer, W.F. Jr, Shea, P.J., Valentine, B., Weatherspoon, C.P. and Lisle, T.E. (technical coordinators). *Proceedings of the Symposium on the Ecology and Management of Dead Wood in Western Forests*. Albany, California. Pacific Southwest Research Station, Forest Service, US Department of Agriculture. General Technical Report PSW-GTR-181, pp. 11–19.

Butler, J.E., Rose, F. and Green, E.E. (2001) Ancient trees – icons of our most important wooded landscapes in Europe. In: Read, H., Forfang, A.S., Marciau, R., Paltto, H., Andersson, L. and Tardy, B. (eds) *Tools for Preserving Woodland Biodiversity*, EU: Naconex, pp. 20–26.

Castellano, M.A., Smith, J.E., O'Dell, T., Cazares, E. and Nugent, S. (1999) *Handbook to Strategy 1 Fungal Taxa from the Northwest Forest Plan*. US Department of Agriculture General Technical Report PNW-GTR-476.

Harding, P.T. and Alexander, K.N.A. (1994) The use of saproxylic invertebrates in the selection and evaluation of areas of relic forest in pasture-woodlands. In: Harding, P.T. (ed.) Invertebrates in the landscape: invertebrate recording in site evaluation and countryside monitoring. Proceedings of the National Federation for Biological Recording Annual Conference, Brighton, 1991. *British Journal of Entomology and Natural History* 7(Suppl. 1), 21–26.

Harding, P.T. and Rose, F. (1986) *Pasture-woodlands in Lowland Britain: a Review of their Importance for Wildlife Conservation.* Institute of Terrestrial Ecology, Huntingdon, UK.

Harding, P.T. and Wall, T. (2000) *Moccas: an English Deer Park.* English Nature, Peterborough, UK.

Hodgetts, N.G. (1992) *Guidelines for Selection of Biological SSSIs: Non-vascular Plants.* Joint Nature Conservation Committee, Peterborough, UK.

Humphrey, J., Stevenson, A., Whitfield, P. and Swailes, J. (2002) Life in the Deadwood. Forest Enterprise, Edinburgh.

IUFRO (1997) http://www.NRCan.gc.ca/hypermail/oldgrowth/

Kirby, K.J. and Watkins, C. (eds) (1998) *The Ecological History of European Forests.* CAB International, Wallingford, UK.

Lund, G.K. (2001) http://home.att.net/~gklund/pristine.html

Nitare, J. (ed.) (2000) *Signalarter. Indikatorer På Skyddsvärd Skog. Flora över Kryptogamer.* [Indicator species for assessing the nature conservation value of woodland sites – a flora of selected cryptogams] Jönköping: Skogsstyrelsens Förlag.

Peterken, G.F. (1996) *Natural Woodland Ecology and Conservation in Northern Temperate Regions.* Cambridge University Press, Cambridge, UK.

Peterken, G.F. (2000) *Natural Reserves in English Woodlands.* English Nature Research Reports, No. 384. English Nature, Peterborough, UK.

Rackham, O. (1986) *The History of the Countryside.* Dent, London.

Rackham, O. (1990) *Trees and Woodlands in the British Landscape* (revised edn). Dent, London.

Rose, F. (1974) The epiphytes of oak. In: Morris, M.G. and Perring, F.H. (eds) *The British Oak. Its History and Natural History.* E.W. Classey, Faringdon, UK, pp. 250–273.

Rose, F. (1976) Lichenological indicators of age and environmental continuity in woodlands. In: Brown, D.H., Hawksworth, D.L. and Bailey, R.H. (eds) *Lichenology: Progress and Problems.* Academic Press, London, pp. 279–307.

Rose, F. (1988) Phytogeographical and ecological aspects of *Lobarion* communities in Europe. *Botanical Journal of the Linnaean Society* 96, 69–79.

Rose, F. (1992) Temperate forest management: its effects on bryophytes and lichen floras and habitats. In: Bates, J.W. and Farmer, A.M. (eds) *Bryophytes and Lichens in a Changing Environment.* Oxford University Press, Oxford, UK, pp. 211–233.

Rose, F. (1993) Ancient British woodlands and their epiphytes. *British Wildlife* 5, 83–93.

Salbitano, F. (1988) *Human Influence on Forest Ecosystems Development in Europe.* Pitagora, Bologna, Italy.

Sanderson, N.A. (1996) *Lichen Conservation within the New Forest Timber Inclosures.* Hampshire Wildlife Trust, UK.

Sanderson, N.A. (1998) Veteran trees in Highland Pasture Woodland. In: Smout, C. (ed.) *Scottish Woodland History Discussion Group Notes III.* Scottish Natural Heritage, Edinburgh, pp. 4–11.

Selva, S.B. (1994) Lichen diversity and stand continuity in the northern hardwoods and spruce-fir forests of northern New England and western New Brunswick. *The Bryologist* 97, 424–429.

Speight, M.C.D. (1989) *Saproxylic Invertebrates and their Conservation*. Nature and Environment Series No. 42. Council of Europe, Strasbourg, France.

Tubbs, C.R. (1997) A vision for rural Europe. *British Wildlife* 9, 79–85.

Vera, F.W.M. (2000) *Grazing Ecology and Forest History*. CAB International, Wallingford, UK.

Watling, R. and Ward, S. (2003) *Fungi Naturally Scottish*. Scottish Natural Heritage, Edinburgh.

The Use of Dendrochronology to Evaluate Dead Wood Habitats and Management Priorities for the Ancient Oaks of Sherwood Forest

18

C. Watkins,[1] C. Lavers[1] and R. Howard[2]

[1]School of Geography and [2]Nottingham University Tree-ring Dating Laboratory, University of Nottingham, Nottingham, UK

Dead wood is increasingly valued as a nature conservation resource and a source of biodiversity, yet there has been relatively little work on the management of standing ancient trees within woodland. Little is known about how long largely dead trees may remain alive, and how long the dead wood is likely to survive before rotting away. In this chapter we evaluate the use of dendrochronology as a means of providing data on the longevity and date of death of ancient oak trees surviving within the remnants of Sherwood Forest. Previous work indicated that around 35% of the cohort of ancient oaks are entirely dead, and large sections of many of the live oaks are dead. As most of the extant oak trees, both dead and alive, are hollow, there has formerly been little interest in carrying out dendrochronological analysis of the trees. In this chapter, rather than using the tree ring data to estimate the age of trees, or look at fluctuations in the growth rate, we use dendrochronology to estimate the date of death and likely longevity of the surviving ancient trees. This information provides a basis for management strategies to ensure the long-term supply of old oak trees.

18.1 Introduction

Dead wood in ancient trees is increasingly valued as a nature conservation resource and a source of biodiversity (Read, 1996; Alexander, 1998). Considerable knowledge has been gained in recent years about the extent of fallen dead wood (Green and Peterken, 1997; Kirby *et al.*, 1998) but there has been relatively little work on the management of dead wood within standing ancient trees. Remarkably little is known about when trees have died, how long largely dead trees may remain alive, or how long the dead wood is likely to survive before rotting away.

As nearly all of Sherwood's ancient oaks are hollow, there has formerly been little interest in carrying out dendrochronological analysis (the age and germination dates of the trees cannot be determined because their cores are missing). Attention has concentrated instead on the analysis of medieval and early modern building timbers, which are usually intact to the core (Laxton *et al.*, 1995; Laxton, 1997). Such research enabled the East Midland Tree-ring Chronology to be constructed (Laxton and Litton, 1988). In this chapter we use the dendrochronological approach for different purposes. Rather than using the tree ring data to estimate the age or germination dates of trees (although we do this where feasible and instructive), the data are used to estimate the date of death and likely longevity of the cohort of surviving ancient trees. This information is useful in guiding the future management both of the ancient trees, and the younger cohorts of oaks, that will eventually take their place as the prime contributors of dead wood habitats to the forest.

18.2 Study Area

The present study was undertaken in the Buck Gates area of Birklands and Bilhagh Special Area for Conservation (cSAC) (Fig. 18.1). Previous work has shown that around 35% of the cohort of ancient oaks in this area are entirely dead, and large sections of many living oaks are dead (Watkins and Lavers, 1998; Clifton, 2000; Clifton and Kirby, 2000).

The site is important because records indicate that there has been little active management over the past 100 years or so, with the following exceptions.

1. A plantation of oak trees was established in the south-eastern section of the site in the 1890s. The Turkey oaks planted then have now been removed, but over 2000 English oaks survive. It is not known how many old oaks were felled when this plantation was made, but the area has by far the highest concentration of large-diameter rotting stumps on the site, which suggests that many ancient trees may have been lost at this time.
2. A small plantation of pines survives in the north-western section of the site. It is not known whether any ancient trees were removed when this was made. One fallen ancient oak remains within the borders of this plantation.
3. A number of felled oak logs survive within the borders of an adjoining army training camp. Some of these may have been moved into the camp from the Buck Gates site in the post-war years.

Fig. 18.1. Distribution of all ancient oaks in the Buck Gates section of Birklands and Bilhagh (Watkins *et al.*, 2003). The solid circles show the sample trees. The lines show the compartment boundaries and rides.

These factors all suggest that the population of ancient oaks is highly likely to have been reduced during the late 19th and 20th centuries.

18.3 Methods

A stratified random sample of 92 trees was taken from the existing database of 406 ancient trees found in the Buck Gates section of Birklands & Bilhagh cSAC (Watkins and Lavers, 1995; Watkins *et al.*, 2003). The main focus of the study was to establish dates of death, but knowing that living trees can be largely dead, we also wanted to take a sample of cores from living trees to investigate, for example, the length of time that a tree may continue to live after portions of it have begun to die. Thus two-thirds of our survey effort was directed towards sampling dead trees and one-third towards living trees. The sample was thus not representative of all ancient oaks on the site, but within our quotas the samples of dead and living trees were chosen randomly. In addition, four reference samples were taken from large, non-ancient oak trees.

Figure 18.1 shows the outline of the Buck Gates section and the distribution of the 92 randomly sampled trees. All the ancient oaks were numbered with metal tags in the mid-1990s (Watkins and Lavers, 1998; Fig. 18.2); to enable other researchers to relate our results to individual trees on the ground, these unique identifiers will be quoted in the tables and figures that follow.

The working method we developed was to take at least two, sometimes three, cores from each sample tree using standard Haglof incremental corers. Samples were taken at breast height (this is usually the most comfortable and safe height for using the corer) and in sections of trunk that were relatively smooth, free from bosses and not deeply rutted. A few samples were also taken from large branches. Wherever possible, cores were taken at places where there was clear evidence of the presence of sapwood (the layer directly beneath the bark). Some fragments of dead wood that appeared to provide ring evidence to the centre of the tree were also collected. The sample cores were prepared for analysis by sanding and polishing. The growth rings were measured and the

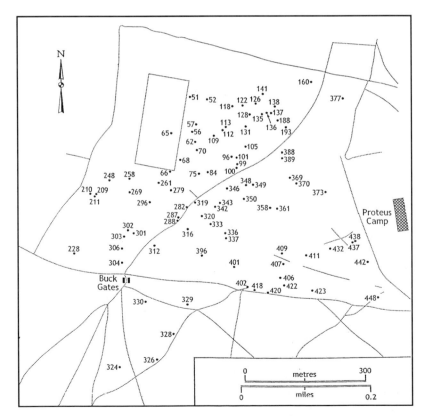

Fig. 18.2. Distribution of sampled trees with their numbers.

growth ring sequences compared with the EMTC (Laxton and Litton, 1988) in order to date the samples.

18.4 Results

Table 18.1 shows data for all 92 ancient trees sampled, together with data for the four reference trees. Note that for six trees there are two sets of results, either because the sample was broken, or because two sample cores have produced useful results. The total number of samples shown in the table, therefore, is 102. It was not possible to make use of 22 samples: 13 were too fragmentary, five had too few rings (samples need at least 55 measurable rings in order to place them in the dendrochronological sequence; Laxton and Litton, 1988), and four had compacted rings. Of the 22 samples that were not measurable, 15 were from dead trees and seven were from living trees. Proportional failure by type of tree was: dead standing and fallen (17%); live trees (20%); 'boats' (fallen trunks with their upper portions rotted away) and stumps (38%). Boats and stumps have a large proportion of their timber in contact with the ground and are most likely to be rotten; it is not surprising, therefore, that these dead wood features are the most difficult to sample effectively.

The sampling procedure produced 80 samples from which accurate measurements of tree rings could be made. Table 18.1 shows the samples dated to the EMTC. Seventy-one of the 80 samples (89%) were successfully matched with this standard sequence (Fig. 18.3). The samples are placed in order of the last known or estimated dated ring. So, for example, the top sample's latest growth ring is 1650, whereas the bottom sample's last growth ring is 2001. Column A shows the position of the first measured ring on each sample in relation to the earliest measured ring of any sample. The earliest measured ring is given the position 000, which is ascribed to the year 1415, taken from tree 336. Column B shows the date of the first measured ring from each sample. Column C shows the total number of measured rings from each sample. The dates in relative years and calendar years are given at the bottom of the figure.

The boundary between heartwood and sapwood is also indicated on Fig. 18.3. Where sapwood was measurable to the bark, the total number of sapwood growth rings is shown, followed by the letter 'C' for 'complete'. Twelve samples had complete sapwood; the number of sapwood rings ranged from 20 to 37, with a mean of 28 growth rings. In six cases it was possible to make an estimate of unmeasured rings which were difficult to measure precisely. These growth rings are indicated separately on the diagram. Twenty-seven of the first 28 samples on the diagram have no identifiable heartwood–sapwood boundary. This makes the identification of the date of death of the sample difficult. The top six samples have a last growth ring of pre-1815. It is likely that these samples have lost a considerable part of their outer growth

Table 18.1. Dendrochronological results for all sampled ancient oaks at Buck Gates, Sherwood, showing number of heartwood and sapwood rings, estimated number of unmeasured heartwood and sapwood rings, and actual or estimated last growth date of tree.

Tree no.	Estimated unmeasured central rings	Relative start position[a]	First measured ring date	Number of measured rings	Relative end position[a]	Last measured ring date	Estimated unmeasured outer rings	Sapwood rings	Estimated last growth date
Ref 1	00	428	1845	158	586	2000	00	28	2000
Ref 2	00	428	1843	158	586	2000	00	20	2000
Ref 3	00	404	1819	98	502	1916	84	20	2000
Ref 4	00	433	1848	152	585	1999	00	25	1999
51	00	–	–	90	–	Undated	00	–	–
52 (pt 1)	00	316	1731	83	399	1813	00	–	–
52 (pt 2)	00	405	1820	84	489	1903	00	–	–
56	00	310	1725	112	422	1836	00	35	1836
57	Short core, too few rings								
62	00	389	1804	196	585	1999	–	28	1999
65	Fragmentary sample								
66	00	276	1691	202	478	1892	00	H/s	1920
68	00	388	1803	120	508	1922	60	H/s	2000
70	00	399	1814	186	585	1999	00	26	1999
75	Short core, too few rings								
84	00	363	1778	94	457	1871	00	–	–
96	60–80	460	1875	125	585	1999	00	30	1999
99	00	372	1787	70	442	1856	00	–	–
100	00	–	–	67	–	Undated	00	–	–
101	00	340	1755	119	459	1873	00	–	–

105	00	–	–	60	–	Undated	40–50	30	–
109	00	299	1714	224	523	1937	00	H/s	1965
112	00	356	1771	188	544	1958	00	H/s	1986
113	00	184	1599	403	587	2001	00	26	2001
118	00	393	1808	143	536	1950	20	15	1983
122	Compacted rings		–	–	–	–	–	–	–
126	Compacted rings		–	–	–	–	–	–	–
128	00	364	1779	140	504	1918	00	H/s	1946
131	00	401	1816	84	485	1899	00	No h/s	–
135	00	398	1813	112	510	1924	00	H/s	1952
136	00	510	1925	73	583	1997	00	25	1997
137	00	361	1776	88	449	1863	00	No h/s	–
138 (pt 1)	00	289	1704	190	479	1893	30	No h/s	–
138 (pt 2)	00	465	1880	66	531	1945	00	H/s	1973
141	00	160	1575	343	503	1917	00	H/s	1945
160	00	382	1797	110	492	1906	80	40	1986
188	00	–	–	47	–	Undated	00	–	–
193	00	306	1721	151	457	1871	40	H/s	1939
209	00	350	1765	236	586	2000	00	26	2000
210	20	365	1780	129	494	1908	00	H/s	1936
211	Fragmentary core		–	–	–	–	–	–	–
228	Fragmentary core		–	–	–	–	–	–	–
248	00	–	–	48	–	Undated	00	–	–

continued

Table 18.1. *Continued.*

Tree no.	Estimated unmeasured central rings	Relative start position[a]	First measured ring date	Number of measured rings	Relative end position[a]	Last measured ring date	Estimated unmeasured outer rings	Sapwood rings	Estimated last growth date
258	00	389	1804	121	510	1924	00	H/s	1952
261	Fragmentary core		–	–	–	–	–	–	–
269	Short core, too few rings		–	–	–	–	00	–	–
279	00	323	1738	160	483	1897	80	50	1977
282	00	310	1725	156	466	1880	00	No h/s	–
287	Fragmentary core		–	–	–	–	–	–	–
288	Compacted rings		–	–	–	–	–	–	–
296	00	401	1816	186	587	2001	00	37	2001
301	00	359	1774	63	422	1836	00	No h/s	–
302	00	221	1636	251	472	1886	00	H/s	1914
303	00	246	1661	241	487	1901	00	H/s	1929
304	00	273	1688	306	580	1994	00	36	1994
306	20	387	1802	96	483	1897	00	No h/s	–
312	Too few rings		–	–	–	–	–	–	–
316	00	313	1728	198	511	1925	00	H/s	1953
319 (pt 1)	00	297	1712	144	441	1855	00	No h/s	–
319 (pt 2)	10	374	1789	100	474	1888	20	H/s	1936
320	10	274	1689	220	494	1908	00	H/s	1936
324	Too few rings		–	–	–	–	–	–	–
326	00	380	1795	120	500	1914	00	H/s	1942

328	Fragmentary sample		—	—	—	—	—	—	—
329	Fragmentary sample		—	—	—	—	—	—	—
330	Fragmentary sample		—	—	—	—	—	—	—
333	00	226	1641	141	367	1781	00	No h/s	—
336	00	00	1415	489	489	1903	00	H/s	1931
337	Fragmentary core		—	—	—	—	—	—	—
342	00	210	1625	283	493	1907	00	No h/s	—
343	Complicated rings		—	—	—	—	—	—	—
346	Fragmentary sample		—	—	—	—	—	—	—
348	00	421	1836	96	517	1931	00	H/s	1959
349	00	—	—	40	—	Undated	00	—	—
350	00	128	1543	108	236	1650	00	No h/s	—
358	00	335	1750	103	438	1852	00	No h/s	—
361	00	331	1746	92	423	1837	00	No h/s	—
369 (pt 1)	00	208	1623	129	337	1751	00	No h/s	—
368 (pt 2)	00	381	1796	72	483	1897	00	No h/s	—
370	00	381	1796	126	507	1921	00	H/s	1949
373	00	—	—	66	—	Undated	00	—	—
377	00	—	—	43	—	Undated	00	—	—
388	00	342	1757	117	459	1873	00	No h/s	—
389	00	—	—	48	—	Undated	00	—	—
396	00	213	1628	59	272	1686	00	No h/s	—
401	00	392	1807	83	475	1889	00	No h/s	—

continued

Table 18.1. *Continued.*

Tree no.	Estimated unmeasured central rings	Relative start position[a]	First measured ring date	Number of measured rings	Relative end position[a]	Last measured ring date	Estimated unmeasured outer rings	Sapwood rings	Estimated last growth date
402	00	332	1747	150	482	1896	00	H/s	1924
406	Fragmentary sample			–		–	–	–	–
407	Fragmentary sample			–		–	–	–	–
409	00	479	1894	73	552	1966	00	H/s	1994
411	00	351	1766	148	499	1913	00	H/s	1941
418	00	195	1610	283	478	1892	00	H/s	1920
420	00	468	1883	77	545	1959	35	20	1994
422 (pt 1)	00	303	1718	146	449	1863	00	No h/s	–
422 (pt 2)	00	417	1832	62	479	1893	00	No h/s	–
423	00	354	1769	63	417	1831	Inestimable	No h/s	–
432	50	374	1789	150	524	1938	60	35	1998
437	Fragmentary sample			–		–	–	–	–
438 (pt 1)	00	330	1745	60	390	1804	00	No h/s	–
438 (pt 2)	00	394	1809	56	450	1864	00	No h/s	–
442	00	372	1787	71	443	1857	00	No h/s	–
448	00	403	1818	101	504	1918	00	H/s	1946

[a]Relative to earliest ring, relative 00, 1415, on sample 336.
h/s, Heartwood/sapwood boundary; –, no data.

rings. One tree (56) has 35 sapwood rings and the last dated ring is 1836, an impossibility because the tree is still alive and the sample was taken through the bark. This tree has been misdated by the dendrochronological method. The remaining 21 top samples have latest growth rings ranging from 1831 to 1907. None has a heartwood–sapwood boundary, but taking the average number of sapwood rings as 28, they should have a minimum number of 28 years added to their last date. This would change the range of dates of death for these 21 samples to 1859 to 1935. It is probable, we feel, that few if any of the resistant heartwood rings have been eroded, and so this range of dates should be fairly accurate. However, as we do not know how many heartwood rings may have been eroded, the addition of the mean number of sapwood rings is not indicated in the figure.

The bottom 43 trees in the diagram all have heartwood–sapwood boundaries. With these, the addition of the mean sapwood growth ring figure (28) when no sapwood is present provides a useful indication of the likely date of death of the sample (not necessarily the whole tree). Taking this modification into account, the samples have dates of death ranging from 1914 to 2001. When grouped into 20-year periods (Table 18.3), there is an indication that a relatively low number of samples have a date of death for the period 1961–1980. This is indicated by the kink in Fig. 18.3 between trees 316 and 118.

Table 18.2 shows the status of the trees from which the samples were taken. Fifty-eight (63%) were from dead trees and 34 (37%) were from living trees. Of the 58 dead trees, 23 (40%) were standing, 10 (17%) were fallen whole trunks, 13 (22%) were 'boats' and 8 (14%) were stumps. Although 37% of the samples were taken from living trees, this does not mean that the sample was necessarily taken through the bark of the living section of the tree. In four cases samples had to be taken from sections where no bark remained, the usual reason being that the surviving bark was heavily bossed and burred, making any sample taken through the bark unusable.

In addition, samples taken through the bark of living trees did not necessarily include tree rings up to the date of sampling. This is because the bark of trees may cover dead sections of the trunk. Ten samples fell into this category (Table 18.3). In the field we found no simple method of identifying whether we had obtained a core from a live or dead section of the tree when coring through bark. A comparison of Table 18.3 with Fig. 18.3 is instructive. Only three of the samples from the top group of 28 with no heartwood–sapwood boundary come from a tree that is alive. One of these, number 56, is misdated. The other two (423 and 342) have last growth rings of 1831 and 1907. For the 30 samples where definite dates of death have been assigned, 17 (57%) are from dead trees and 13 (43%) from live trees. Thus, while it is possible to estimate the date of death of the section of tree from which individual samples are taken, it is not possible unequivocally to provide a definite estimate of the date of death of the trees themselves, because different sections of the tree will cease growing at different times.

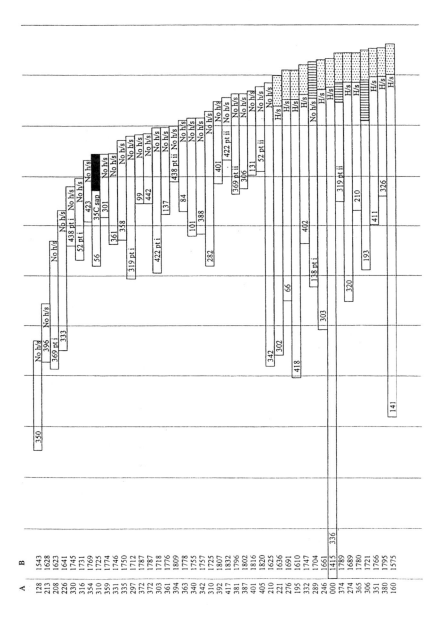

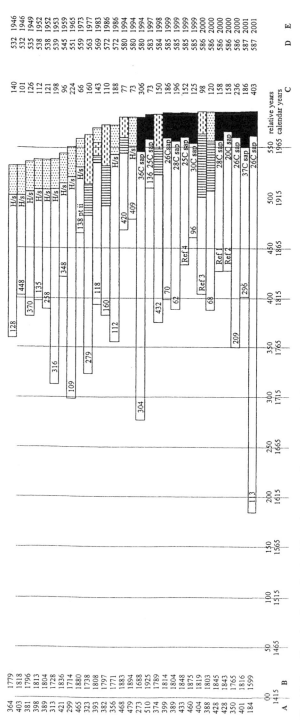

Fig. 18.3. Dendrochronological bar diagram of cross-matched and dated samples, Buck Gates ancient oaks. Shading: black, measured sapwood rings; dotted, mean of 28 sapwood rings added; vertical lines, estimate of heartwood rings from partial data; dashes, estimated sapwood rings where they are difficult to measure. A, This is the position of the first measured ring on the sample in relation to the earliest measured ring of all the sample, 000, 1415, on sample 336. B, Date of first measured ring. C, Number of measured rings in sample. D, This is the position of the estimated or actual last growth ring of the tree where sapwood exists or is estimated, position relative to the earliest ring of all the samples, 000, 1415 on sample 336. Where there is no sapwood or the heartwood/sapwood (h/s) boundary is not present, this is the relative position of the last measured ring date on the sample. E, Where sapwood exists, or is estimated, this is the date of the actual or estimated last growthring of the tree, i.e. the death date. Otherwise, where sapwood or the heartwood/sapwood (h/s) boundary is not present, this is the date of the last measured ring on the sample.

Table 18.2.   Tree status and dendrochronological results.

Tree no.	Status	Form[a]	Branch order[b]	Core position[c]	First ring date	Last ring date	Estimated last growth date	Notes
Ref 1	Alive	Standing	T	Bark	1845	2000	2000	Non-ancient tree
Ref 2	Alive	Standing	T	Bark	1843	2000	2000	Non-ancient tree
Ref 3	Alive	Standing	T	Bark	1819	1916	2000	Non-ancient tree
Ref 4	Alive	Standing	T	Bark	1848	1999	1999	Non-ancient tree
51	Dead	Standing	0	Heartwood	1731	1903	No estimate	Does note date
52	Dead	Standing	2	Heartwood	1725	1836	1836	No heart/sap boundary
56	Alive	Standing	T	Bark			No estimate	Short core
57	Alive	Standing	T	Bark			No estimate	
62	Alive	Standing	T	Bark			No estimate	Fragmented core
65	Dead	Fallen whole	0	Heartwood	1804	1892	No estimate	
66	Alive	Standing	T	Heartwood	1691	1922	1920	
68	Alive	Standing	T	Bark	1803	1999	2000	
70	Alive	Standing	T	Bark	1814	1999	1999	
75	Alive	Standing	T	Bark			No estimate	Short core
84	Dead	Standing	1	Bark	1778	1871	No estimate	No heart/sap boundary
96	Alive	Standing	T	Bark	1875	1999	1999	
99	Dead	Standing	2	Heartwood	1787	1856	No estimate	No heart/sap boundary
100	Alive	Standing	T	Bark			No estimate	Does note date
101	Dead	Fallen whole	1	Heartwood	1755	1873	No estimate	No heart/sap boundary
105	Alive	Standing	T	Bark			No estimate	Does note date
109	Dead	Fallen boat	3	Heartwood	1714	1937	1965	
112	Dead	Standing	2	Heartwood	1771	1958	1986	
113	Alive	Standing	T	Bark	1599	2001	2001	

118	Dead	Fallen boat	2	Heartwood	1808	1950	1983	
122	Dead	Standing	2	Bark			No estimate	Compacted rings
126	Alive	Standing	T	Bark			No estimate	Compacted rings
128	Alive	Standing	T	Bark			1946	
131	Dead	Fallen whole	2	Heartwood	1779	1918	No estimate	No heart/sap boundary
135	Alive	Standing	T	Bark	1816	1899	1952	
136	Alive	Standing	T	Bark	1813	1924	1997	
137	Dead	Fallen boat	1	Heartwood	1925	1997	No estimate	No heart/sap boundary
138	Alive	Standing	T	Bark	1776	1863	1973	
141	Dead	Standing	2	Bark	1704	1945	1945	
160	Alive	Standing	T	Bark	1575	1917	1986	
188	Alive	Standing	T	Bark	1797	1906	No estimate	Does note date
193	Alive	Standing	T	Heartwood	1721	1871	1939	
209	Alive	Standing	T	Bark	1765	2000	2000	
210	Dead	Standing	1	Heartwood	1780	1908	1936	
211	Alive	Standing	T	Bark			No estimate	Fragmented core
228	Alive	Standing	T	Bark			No estimate	Fragmented core
248	Dead	Standing	3	Heartwood			No estimate	Does note date
258	Dead	Standing	2	Heartwood	1804	1924	1952	Fragmented core
261	Dead	Fallen boat	0	Heartwood			No estimate	Short core
269	Dead	Standing	2	Heartwood			No estimate	
279	Alive	Standing	T	Bark			1977	
282	Dead	Standing	2	Heartwood	1738	1897	No estimate	No heart/sap boundary
287	Dead	Fallen boat	0	Heartwood	1725	1880	No estimate	Fragmented core
288	Alive	Standing	T	Bark			No estimate	
296	Alive	Standing	T	Bark	1816	2001	2001	Compacted rings

continued

Table 18.2. *Continued.*

Tree no.	Status	Form[a]	Branch order[b]	Core position[c]	First ring date	Last ring date	Estimated last growth date	Notes
301	Dead	Fallen boat	0	Heartwood	1774	1836	No estimate	No heart/sap boundary
302	Dead	Standing	1	Heartwood	1636	1886	1914	
303	Alive	Standing	T	Bark	1661	1901	1929	
304	Alive	Standing	T	Bark	1688	1994	1994	
306	Dead	Standing	2	Heartwood	1802	1897	No estimate	No heart/sap boundary
312	Dead	Standing	1	Heartwood			No estimate	Short core
316	Dead	Standing	5	Bark	1728	1925	1953	
319	Dead	Standing	2	Bark	1712	1888	1936	
320	Alive	Standing	T	Bark	1689	1908	1936	
324	Alive	Standing	T	Heartwood			No estimate	Short core
326	Alive	Standing	T	Heartwood	1795	1914	1942	
328	Dead	Stump	0	Heartwood			No estimate	Fragmented core
329	Dead	Stump	1	Heartwood			No estimate	Fragmented core
330	Dead	Fallen boat	0	Heartwood			No estimate	Fragmented core
333	Dead	Standing	1	Heartwood	1641	1781	No estimate	No heart/sap boundary
336	Dead	Fallen whole	0	Heartwood	1415	1903	1931	
337	Dead	Standing	0	Heartwood			No estimate	Fragmented core
342	Alive	Standing	T	Bark	1625	1907	No estimate	No heart/sap boundary
343	Dead	Fallen whole	2	Heartwood			No estimate	Compacted rings
346	Dead	Standing	2	Heartwood			No estimate	Fragmented core
348	Dead	Standing	2	Heartwood	1836	1931	1959	Does note date
349	Dead	Fallen whole	1	Heartwood			No estimate	
350	Dead	Stump	0	Heartwood	1543	1650	No estimate	No heart/sap boundary

358	Dead	Fallen boat	1	Heartwood	1750	1852	No estimate	No heart/sap boundary
361	Dead	Fallen whole	2	Heartwood	1746	1837	No estimate	No heart/sap boundary
369	Dead	Standing	1	Heartwood	1623	1897	No estimate	No heart/sap boundary
370	Dead	Standing	3	Bark	1796	1921	1949	
373	Dead	Standing	2	Heartwood			No estimate	Does note date
377	Dead	Stump	0	Heartwood			No estimate	Does note date
388	Dead	Fallen boat	0	Heartwood	1757	1873	No estimate	No heart/sap boundary
389	Dead	Stump	0	Heartwood			No estimate	Does note date
396	Dead	Fallen whole	1	Heartwood	1628	1686	No estimate	No heart/sap boundary
401	Dead	Fallen whole	0	Heartwood	1807	1889	No estimate	No heart/sap boundary
402	Dead	Fallen boat	1	Heartwood	1747	1896	1924	
406	Dead	Stump	0	Heartwood			No estimate	Fragmented core
407	Dead	Fallen boat	1	Heartwood			No estimate	Fragmented core
409	Dead	Standing	2	Bark	1894	1966	1994	
411	Dead	Stump	0	Heartwood	1766	1913	1941	
418	Dead	Stump	0	Heartwood	1610	1892	1920	
420	Alive	Standing	T	Bark	1883	1959	1994	
422	Dead	Fallen whole	0	Heartwood	1718	1893	No estimate	No heart/sap boundary
423	Alive	Standing	T	Bark	1769	1831	No estimate	No heart/sap boundary
432	Alive	Standing	T	Bark	1789	1938	1998	
437	Dead	Fallen boat	0	Heartwood			No estimate	Fragmented core
438	Dead	Standing	0	Heartwood	1745	1864	No estimate	No heart/sap boundary
442	Dead	Standing	1	Heartwood	1787	1857	No estimate	No heart/sap boundary
448	Dead	Fallen boat	0	Heartwood	1818	1918	1946	

a 'Fallen boat', as opposed to 'fallen whole', indicates a hollow, prone trunk with its upper portion rotted away.
b T, twigs present; 1, first-order branches only present; 2, first- and second-order branches present, etc.
c Whether core was taken through bark or straight into heartwood.

Table 18.3. Deaths of oak trees in 20-year periods during the 20th century.

Last growth ring	−1920	1921–1940	1940–1960	1961–1980	1981–1997	Alive
Number of samples	3	8	10	3	7	12

The inner dates shown in Fig. 18.3 indicate how long we might expect dead wood to survive within the trees. The oldest piece of wood found was produced in the year 1415. Thirty-one (43%) of the trees contain wood dated accurately to 1765 or earlier; 57 (80%) of the trees contain timber dated back to 1815 or earlier.

It is interesting that the number of surviving growth rings in trees that have no identifiable heartwood–sapwood boundary (Fig. 18.3) is significantly lower than for trees with a clear boundary. Samples with no heartwood–sapwood boundary ($n = 22$) have a mean number of 119 growth rings; those with a heartwood–sapwood boundary ($n = 39$) have a mean of 181 (Mann–Whitney U = 199, $P < 0.01$). Our observations suggest that remnant cylinders continue to rot away predominantly from the inside towards the outer heartwood, although erosion of exposed outer rings may also occur.

18.5 Discussion and Management Implications

We have used dendrochronological analysis of dead and living oak trees to estimate the date of death and likely longevity of the surviving ancient trees in Sherwood Forest. This information is of importance in determining the future management of the remaining fragments of Sherwood Forest, in particular for providing a sound basis for the development of management strategies to ensure the long-term supply of dead wood.

This chapter confirms that:

1. It is possible to extract samples using standard wood corers from ancient dead and living oaks.
2. The rings of what were often very slowly growing oak trees are decipherable in the dendrochronology laboratory.
3. It is possible to place the ring sequences within the established East Midlands Tree-ring Chronology.
4. It is often possible to distinguish the sapwood/heartwood boundary in cores (when sapwood is not present), and thus make an estimate of the date of death for individual samples from trees.

The analysis of the ancient trees at Sherwood provides no clear evidence that any particular factor caused the death of many trees in a short period of time. The cohort appears to be fading away 'naturally' at a steady rate. Analysis of the date of death of samples (not necessarily trees) shows that an average of about six samples is dying off in each 20-year period.

A possible reason for the high death rates in the middle years of the 20th century (Table 18.3) is felling, associated with wartime activity such as tank manoeuvres. Another reason frequently given for death of oak trees is air pollution, but these figures do not strongly support the idea of a significant impact of air pollution on the trees in mid-century.

Field observations suggest that the following sequence of senescence is probably typical for the ancient oaks of Sherwood:

1. Heart rot attacks the tree's core, leaving it hollow (for modes of initiation and mechanisms of heart rot, see Rayner and Body, 1988). When the rate of heart rot exceeds the rate at which new wood is added to the growing margin, the total amount of wood present in the trunk begins to decline.

2. Sections of trees die at different times. Some sections of trunk may be entirely dead, with no actively growing cambium, while other sections of trunk are still alive.

3. When a section of the trunk dies, its bark and sapwood disintegrate relatively quickly, leaving the youngest layer of heartwood exposed.

4. While there may be some erosion of this outer layer of heartwood once the overlying bark and sapwood have disappeared, heart rot eliminating material from the interior of the tree is likely to be by far the most important cause of wood loss from the trunk.

5. Eventually the whole tree dies, most of its branches, bark and sapwood disappear, and a whole or partial cylinder of heartwood remains.

6. The trunk continues to rot from the inside towards the outer layer of heartwood. If lower sections of the trunk rot faster than upper sections, the trunk eventually falls; if the reverse happens, the standing trunk above ground gradually decays until only a stump remains.

An issue of major conservation concern in Sherwood and other forests containing ancient trees is the long-term continuity of dead wood habitats. Of the 58 dead trees sampled in this study, we are confident that 52 died after 1850. (The earliest last growth ring among these samples, with the addition of 28 rings for missing sapwood, dates to 1859; it is possible that some heartwood rings are missing from samples where the heartwood–sapwood boundary is not visible, but addition of extra rings to account for this would simply push the estimated date of death forward in time.) The remaining six trees may also have died after 1850 if, as we suspect, their early apparent death dates are the result of erosion of outer rings, but of this we cannot be certain. Nevertheless, remembering that the date of the last ring on any core (plus 28 for missing sapwood if necessary) indicates the *earliest* possible date of death of the tree, it appears that the great majority of dead trees still detectable in our study area died within the past 150 years. This implies that trees which died more than 150 years ago are likely to be completely rotted, or at least in an advanced state of decay. In other words, our best estimate of the time it takes for an ancient oak to rot away in Sherwood after the death of all or some portion of it, is 150 years or less. We are aware that such estimates could be of strictly limited value due to high variability in decay rates. Some oaks in our study area are still alive despite losing 90% of their volume to rot. When such trees finally die the little wood that remains is likely to rot away quite quickly. Conversely, tree 336 died between the First and Second World Wars at the age of around 530, was solid to the core on death, and is still solid today. However, despite these and many other complicating factors, to our knowledge our estimate is the best currently available for Sherwood Forest and the only one based on an empirical study.

Sixty-five per cent of the ancient oaks in our study area are alive. Should they all die in a few years, our results suggest that they might continue to contribute dead wood habitats to the forest for a century or so. The other major cohort of trees in the area is 110 years old (known from forestry records and tree-ring counts of recently felled Turkey oaks); 150 years hence these trees will be 260 years old. Tree 336 in our study was around 530 when it died in the first half of the 20th century, and tree-ring counts from other parts of Sherwood suggest that the ancient trees are mostly between 400 and 600 years old (Clifton, 2000). Were the current crop of ancient trees to die off very quickly, therefore, they could conceivably rot away before the upcoming cohort reaches the state of over-maturity characteristic of the current veteran population.

Our results emphasize the need to preserve and maintain the deadwood habitats in Sherwood and to ensure that the remaining living ancient trees are fully protected so that their deaths are not hastened by human action. Protective management over the past few years has included the removal of conifers and introduced broadleaved trees, the reintroduction of grazing and the control of bracken. The removal of conifers immediately surrounding the ancient trees appears to have a beneficial effect on the survival of ancient oaks. This policy should continue and be extended over the whole of the site of special scientific interest. The removal of encroaching birch trees which compete with the ancient oaks may also be a wise precaution. The monitoring of the ancient oaks and the assessment of the effects of different management prescriptions needs to be continued.

References

Alexander, K.N.A. (1998) The links between forest history and biodiversity: the invertebrate fauna of ancient pasture-woodlands in Britain and its conservation. In: Kirby, K. and Watkins, C. (eds) *The Ecological History of European Forests*. CAB International, Wallingford, UK, pp. 73–80.

Clifton, S. (2000) The status of Sherwood's ancient oaks. *Transactions of the Thoroton Society* 104, 51–63.

Clifton, S. and Kirby, K. (2000). The veteran trees of Birklands and Bilhagh, Sherwood Forest, Nottinghamshire. *English Nature Research Report* No. 316.

Green, P. and Peterken, G.F. (1997) Variation in the amount of dead wood in the woodlands of the lower Wye Valley, UK in relation to the intensity of management. *Forest Ecology and Management* 98, 229–238.

Kirby, K.J., Reid, C.M., Thomas, R.C. and Goldsmith, F.B. (1998) Preliminary estimates of fallen dead wood and standing dead trees in managed and unmanaged forests in Britain. *Journal of Applied Ecology* 35, 148–155.

Laxton, R. (1997) A 13th century crisis in the Royal Forest of Sherwood in Nottinghamshire. *Transactions of the Thoroton Society* 101, 73–98.

Laxton, R. and Litton, C. (1988) *An East Midlands Master Tree-Ring Chronology and its use for Dating Vernacular Buildings.* Department of Classics and Archaeology, University of Nottingham Monograph III.

Laxton, R., Litton, C. and Howard, R. (1995) Nottinghamshire houses dated by dendrochronology. *Transactions of the Thoroton Society* 99, 45–54.

Lott, D. (1999) A comparison of saproxylic beetle assemblages occurring under two different management regimes in Sherwood Forest. *Naturalist* 124, 67–74.

Mitchell, A. (1996) *Trees of Britain.* Collins, London.

Rayner, A. and Boddy, L. (1988) *Fungal Decomposition of Wood: Its Biology and Ecology.* John Wiley & Sons, Chichester, UK.

Read, H.J. (1996) *Pollard and Veteran Tree Management II.* Corporation of London, Burnham Beeches.

Watkins, C. (1998) 'A solemn and gloomy umbrage': changing interpretations of the ancient oaks of Sherwood Forest. In: Watkins, C. (ed.) *European Woods and Forests. Studies in Cultural History.* CAB International, Wallingford, UK, pp. 93–114.

Watkins, C. and Lavers, C. (1995) *Veteran Oak Tree Survey, Birklands and Bilhaugh.* Report for English Nature.

Watkins, C. and Lavers, C. (1998) Losing one's head in Sherwood Forest: the conservation of the ancient oaks. In: Atherden, M. and Butlin, R. (eds) *Woodland in the Landscape: Past and Future Perspectives.* Leeds, UK, pp. 140–152.

Watkins, C., Lavers, C. and Howard, R. (2003) *Veteran Tree Management and Dendrochronology, Birklands and Bilhaugh cSAC, Nottinghamshire.* English Nature Research Reports 489, English Nature, Peterborough, UK.

Forest Regulations in the USA: Evolving Standards for Conserving Forest Biodiversity in the Past 300 Years

19

M.J. Mortimer

Virginia Polytechnic Institute, State University, Department of Forestry, Blackburg, Virginia, USA

Changes in the focus and manner of forest regulation in the USA demonstrate a marked shift in public priorities over the past 300 years. Biodiversity, and the protection or maintenance thereof, has until recently lacked emphasis in the public regulation of private forestlands. However, private forestland in the USA is a crucial component in any effort to conserve forest biodiversity. This chapter suggests that though there may be new-found emphases on forest biodiversity and other non-traditional forest management goals on private forestland, regulating for such goals moves ever nearer to conflicts with constitutionally guaranteed private property rights. Avoiding those conflicts, by considering alternatives to biodiversity regulation, will primarily require incentives or market-based approaches. Over-reliance upon regulatory mechanisms runs a high risk of producing undesirable political and management results, including conversion of forestland to alternative non-forest uses.

19.1 Background

Forest ownership in the USA is a fragmented and complex affair. Ownership of the 302 million ha of US forests[1] is both public and private, and amongst those

---

[1] Forestland is defined in pertinent part by the USDA RPA as 'land at least 10 percent stocked by forest trees of any size, including land that formerly had such tree cover and that will be naturally or artificially regenerated'. The minimum size for this classification is 1 acre.

©CAB International 2004. *Forest Biodiversity: Lessons from History for Conservation*
(eds O. Honnay, K. Verheyen, B. Bossuyt and M. Hermy)

two broad ownerships groups are subsets of owners (Fig. 19.1). Public forests
are primarily of two types, those managed by the federal government,
including approximately 78 million ha of national forests, and those managed
by the state governments. While national forest management is largely homogeneous, due to the centralized federal management legislation, the 50 states
manage their respective public forestlands for a number of goals, including
long-term income generation (Souder and Fairfax, 1996). Although American
Indian tribes also manage various forestlands, the regulation of tribal lands is a
detailed subject unto itself and beyond the scope of this chapter.

Nearly 70% of pre-European settlement forests (*c.* 1630) are estimated to
exist currently (USDA, 2002a). While current net volumes of forest are largely
stable, conversion of forestland to agriculture, or vice versa, and the more
recent conversions of forestland to urban uses have occurred throughout the
history of the nation. Some 9.9 million private landowners hold approximately
144 million ha, or nearly 71% of the nation's timberland[2] (USDA, 1996,
2002b). Production from these lands accounted for 89% of US timber
harvested in 1996 (USDA, 1997), and is likely to increase as national forest
harvest levels continue to decline. Private landowners are of two primary
types: industrial and non-industrial. The former are categorized by large,
corporate land holdings, managed primarily to provide a continuous supply
of timber to satisfy commercial sawmills, paper mills, or other industrial
operations. Plantation forestry is often characteristic of industrial operations.
Non-industrial private landowners (NIPFs) represent a broad spectrum of acre-
ages, management goals and forest types (Birch, 1996). The unifying theme
among NIPF lands is their general dissociation with industrial operations,
though they, too, regularly provide wood fibre for industrial operations.

Private forestlands in the US are notably important:

The potential of nonfederal forestlands to contribute to the maintenance of
biodiversity is great, given their extent, variety, potential management flexibility,

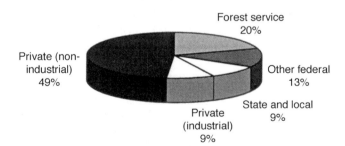

Fig. 19.1.   US forest ownership (Best and Wayburn, 2001).

---

[2] Timberland is defined in pertinent part by the USDA RPA as 'forest land that is producing or
is capable of producing crops of industrial wood and not withdrawn from timber utilization
by statute or administrative regulation'. Timberland must be able of producing at least 20
cubic feet per acre per year of industrial wood in natural stands.

and that they are the primary forest category subject to conversion to non-forest uses

(National Research Council, 1998, p. 59)

The public/private distinction is also reflected in the manner in which these forests are regulated. Management of federal public lands is the exclusive responsibility of the US Congress, and designated management agencies such as the Forest Service or Bureau of Land Management. While federal law does not typically regulate state and private lands directly, federal statutes do indirectly impact management. Laws such as the Federal Water Pollution Control Act (Clean Water Act) and the Endangered Species Act (ESA) have had profound impacts on permitted forestry activities, regardless of the underlying ownership.

State public lands are analogous to federal public lands as their management is determined by state statute, or by combinations of state and federal statutes. Private land is regulated primarily by the states through statutes and administrative regulations, although federal laws, dealing primarily with water quality and endangered species, may also exert indirect control.

19.2 The Nature of Regulation

Forest regulatory trends are primarily a private forestland issue. While the uses of public forestland, particularly timber harvesting, are certainly subject to management restrictions, those restrictions are not in the form of *regulations*, but are instead encompassed in the legal *management objectives* for the land. Private forestlands, on the contrary, reflect no particular management goal. Private forestlands are managed in accordance with the desires of the individual landowners, and are therefore subject to changes over time.

Private property rights, or more specifically private real property rights, are a hallmark of US jurisprudence and culture. Reverence for private property is embodied in the federal constitution and reflected in state constitutions. The history of this paradigm is long and interesting, but can be encapsulated in the 18th century writings of the English legal scholar William Blackstone:

> So great moreover is the regard of the law for private property, that it will not authorize the last violation of it; no, not even for the public good of the whole community . . . In vain may it be argued, that the good of the individual ought to yield to that of the community; for it would be dangerous to allow any private man, or even any public tribunal, to be judge of this common good, and to decide whether it be expedient or no

(Blackstone, 1979, p. 135)

Blackstone's view was influenced by centuries of regard for real property. The early Germanic inhabitants of the British Isles held land ownership as integral to full freedom (Green, 1877). It is hardly surprising, then, that restrictions on the unfettered use of private real property have been treated so seriously in US jurisprudence.

Regulation of private forestland is primarily accomplished by use of the various states' police powers, subject to certain federal and state constitutional limitations (Hickman and Hickman, 1997). Police powers are typically those inherent powers exercised by a sovereign government, designed to protect the public's health, safety and welfare. Some argue that in deciphering questions of the legal exercise of the police power, an anti-nuisance test is the means to avoid overbroad or unchecked use of this power (Epstein, 1985). This test, addressed later, is particularly salient to categorizing forestry-related regulations.

While a state's police power is indeed broad, it is not without bounds. Because the police power has the potential to affect citizens' rights, particularly property rights, there are both federal and state safeguards against overreaching by the sovereign. While the police power provides a government with the ability to issue regulations that affect the rights and values of private property, the various constitutions prohibit the taking of the entire value of the property without compensation to the landowner. In essence, a sovereign may not disguise eminent domain as a police power exercise.

Although the federal constitution provides for compensation to landowners in cases of total forfeitures of the value of the real property, and arguably in cases of less than total forfeitures, state constitutions often explicitly provide for cases of less than total forfeiture. *Damage* to the value of real property by virtue of a forest regulation can therefore be afforded greater protections at the state level than at the federal (Mortimer *et al.*, 2003). This coincides with the state primacy in forest regulatory authority.

The diverse nature of modern US forestry regulations virtually assures that regulatory interaction with private property rights will serve as a source of conflict. This potential is well recognized (Society of American Foresters, 2002). The State of Utah, for example, retains a private property ombudsman responsible for addressing private property conflicts at the state and local level. The legal and cultural importance of real property in the US should not be underestimated, though recognizing that legal and cultural attitudes will vary by state and region.

The spectrum of forest management mechanisms available to influence US forest landowners is quite broad, with regulation at the far end (Fig. 19.2). Choices of particular mechanisms are largely a result of political struggles at the state level.

19.3 The Nature of Biodiversity

Assorted research has attempted to design systematic means for assessing the state of biodiversity in US forests. As a signatory to the Montreal Process

Indicators,[3] the federal Forest Service embarked on establishing a number of sustainable forestry criteria, the first of which is *conservation of biological diversity* (USDA, 2002a). The proposed indicators within Criterion 1 include: the extent of forest type relative to total forest areas; extent of area by forest type and age class; extent of forest type in protected areas categories (and defined by age class); fragmentation of forest types; number of forest-dependent species; the status of forest-dependent species (rare, threatened, endangered or extinct); the number of forest species that occupy a small portion of the former range; and the population levels of representative species from diverse habitats monitored across their range (USDA, 2002a).

Others have identified a compositional component, a structural component and a functional component, all of which can be assessed and monitored to an extent by assessment of existing vegetative structure (Pregitzer *et al.*, 2001).

Finally, a recent multidisciplinary report identified six indicators to describe the biological condition of US forests: percentage of forest-dwelling species at different levels of risk of extinction; percentage of plant cover in non-native forest; forest age structure; acres of forest affected each year by fires, insects and tree pathogens; fire frequency; and area of forest occupied by forest types that have suffered significant decline since settlement (Heinz Center, 2002).

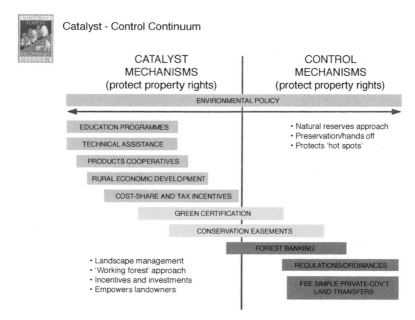

Fig. 19.2.   Private forestland policy tools (courtesy: Dylan Jenkins, 2002).

---

[3] The USA and 11 other countries on five continents signed the Montreal Process. These 12 countries contain 60% of the world's forests (USDA, 2002a).

Common to all these approaches are assessments of the attributes and quantity of various forest types, species, and age classes. Also common is a concern for threats to particular animal and plant species. It is, of course, a far cry from measuring, mapping, and monitoring these conditions to enacting regulations to address possible shortfalls.

Recognizing that sustainable management has been an historic hallmark of professional forestry (Wiersum, 1995), today's questions must be: what shall be the goal of future forms of sustainable management; and how will those goals be established? For example, regulating for biodiversity sustenance may require land-use controls incorporating prescriptions on the number, species, and geospatial locale of trees in timber harvesting operations. While the historic goals of sustainable forest management may not have often crossed constitutional bounds, integration and regulation of the various biodiversity criteria and indicators may imply crossing that line.

19.4 Public Lands

As discussed above, public lands in the US are not the subjects of forest regulation *per se*, but rather reflect evolving management philosophies embodied in legislation and administrative rules. It is important for the purpose of this chapter to note that sustainable management has always been a hallmark of the national forests. Examples include the continuous production of wood and water,[4] the sustained production of wood, water, forage, recreation, and fish and wildlife,[5] biodiversity,[6] and ecological sustainability.[7] None the less, the political vagaries of federal land management are not transferable to private land, as the diverse ownership of private lands makes centralized management prescriptions infeasible.

19.5 Private Lands

Private forestland in the USA has, to varying extents, been subjected to regulation for at least the past 50 years. Prior to that time, very little private land regulation existed, and that which did exist was little concerned with biodiversity. The Plymouth Colony's timber export restrictions (1626), the Penn's Colony restrictions on clearcutting and tree retention (1681), and the

---

[4] Organic Administration Act of 1897, 16 USC 473–475, 477–482, 551.
[5] Multiple-use Sustained-yield Act of 1960, 16 USC 528–531.
[6] National Forest Management Act of 1976, 16 USC 472a, 476, 500, 513–516, 518, 528, 576b, 1600–1602, 1604, 1606, 1608–1614.
[7] See *Sustaining the People's Lands, Recommendations for Stewardship of the National Forests and Grasslands into the Next Century*. Committee of Scientists, US Department of Agriculture, Washington DC, and the resultant federal regulations.

Broad Arrow policy embedded in the Massachusetts Colony Charter (1691), were all primarily concerned with the commercial values of timber.

Until the turn of the 19th century, very little forest regulation was recognizable, for even then a number of states were still coalescing from the US territorial acquisitions of the 1800s. For all intents and purposes, US forestry regulations throughout the 17th, 18th and 19th centuries could barely be considered fledgling, as US forests were still largely viewed solely as exploitable resources or as impediments to progress (Dana and Fairfax, 1980). The earliest forest regulations stemmed from perceived over-harvesting and largely addressed site regeneration and future timber productivity (Laitos, 2002). These early laws did not explicitly reflect a concern for biodiversity, ecosystems or current notions of sustainable forestry. However, to the extent that sustainable forest regeneration and growth is a reflection of forest health, state laws and policies addressing those specific issues could be considered embryonic proxies for grander concepts of biodiversity (Zeide, 2001).

While early attempts were made to regulate private forestlands by federal legislation, the efforts were short-lived. During the 1930s and 1940s, 13 states enacted forest management laws in an effort to preclude threats of federal regulation, and in response to the National Industry Recovery Act of 1933 and its associated Lumber Code (Cubbage, 1997). All 13 sets of regulations focused on adequate regeneration after harvest, implicitly addressing biodiversity to the extent that harvested stands would be regenerated. However, no provision was made for an explicit or comprehensive treatment of biodiversity. The US Supreme Court, on various constitutional grounds, including violation of the state's reserved powers under the Tenth Amendment, invalidated the Lumber Code in 1935. However, several states that enacted forest practices acts while the Lumber Code was in effect, including Washington and Oregon, have continued to mandate some of the most rigorous forest practices standards in the nation.

The aforementioned time period is considered the first of three phases in US forest practices regulation (Ellefson et al., 1997). It could best be described as a simplistic attempt at regulating private forestland, and then only in the wake of a federal regulatory threat. The second phase of private forest regulation occurred again in response to federal actions, in this case to sweeping environmental legislation enacted during the 1960s and 1970s. These forest practices laws are characterized by the federal mandates they sought to satisfy. The enactment of the Clean Water Act (1972), the Clean Air Act (1955, amended in 1970), the Coastal Zone Management Act (1972), and the Endangered Species Act (1973) particularly, required states to take actions to comply with federally established standards and mandates. Most importantly, water quality and threats to endangered species dictated that timber harvesting operations should not unnecessarily impair state water bodies, nor result in the 'take' of an endangered species. The focus shifted from timber to timber-related resources.

Characterizing the most recent or modern phase of US private forest regulation is an expansion of both the substantive regulatory areas – wildlife, fisheries, soil productivity, recreation and aesthetics – and an expansion of the regulating authorities, as local governments have come to play an increasing role in the past decade (USDA, 2002b). The proliferation of local forestry ordinances is a hallmark of this *modern* regulatory phase.

State forest practices acts typically direct state departments of forestry to establish regulations designed to implement the laws. Embedded within those administrative regulations are often the detailed prohibitions or prescriptions for various forest management activities. However, even with the expansion of regulatory specialties and regulating bodies, the explicit preservation or enhancement of biodiversity has remained largely absent.

Those relatively rare state forest practices acts explicitly addressing species concerns are none the less designed with particular species in mind – the spotted owl and marbled murrelet in Washington (Washington Department of Natural Resources, 2000) and California (California Division of Forestry and Fire Protection, 2001) for example. They do not reflect an overarching policy of biodiversity conservation or enhancement. Even the most recently enacted state forest practices act fails even to mention biodiversity (Utah Code, 2001).

While current laws may prohibit private forest landowner actions that may imperil an enumerated or particularly sensitive species, no US state has opted to embrace biodiversity holistically (Cubbage, 1997). Discrete components and indirect benefits of biodiversity are indeed embedded within existing codes, but current public or scientific concerns over forest biodiversity remain largely unfulfilled.

19.6 Impediments and Opportunities

The seeming reluctance for states and localities to enact legislation or to craft administrative regulations directly addressing biodiversity may at first appear perplexing. Consider that state governments, by and large, have far more power to enact legislation than does the federal government, and that, historically, state forest regulations have generally been legally upheld (Warren, 1950). And while some would argue that the US public national forests are moving in the direction of managed biodiversity (Thomas, 2000) and so too some state public forests (Montana Department of Natural Resources and Conservation, 2001), private forest land regulations are not mirroring that trend.

Explaining the seeming disparity between public land and private land reactions to questions of biodiversity is, in all likelihood, multifactorial. First, management objectives for federal public lands often carry no requirement for the generation of income. Thus, restriction of forest management activities for the sake of biodiversity is politically and legally more acceptable than on private lands. The admittedly difficult task the US Congress and President face in seeking new national forest legislation pales in comparison with the momentum necessary to promulgate similar regulations affecting private lands.

Secondly, questions of the definition, the nature, and the role of biodiversity remain far from being settled conclusively. While some see biodiversity as the foundation of a new management ethos (Marcot, 1997), others see it as an unscientific concept riddled with flaws (Fitzsimmons, 1999). These ideological debates are not easily resolved on the vast number of private forestland tracts, where individuals own the land and make the management decisions.

There also exists a tension between the various forest regulatory mechanisms available within the USA. While some infringe very little on the prerogatives of individual landowners, others set standards or prescribe prohibited practices (Fig. 19.2). These 'control mechanisms' often elicit powerfully negative responses from private landowners, while remaining a powerful tool for environmental non-governmental interest groups interested in biodiversity protection policy (Cubbage, 1997).

Finally, constitutional safeguards have the potential to clash with efforts to aggressively promote biodiversity laws and regulations. A tremendous volume of legal literature has been crafted detailing the intricacies of US constitutional protections of real property (Meltz *et al.*, 1999). By and large, the focus has been on the federal constitution and its prohibition against a complete *forfeiture* of the value of the subject property by virtue of the offending regulation. That situation is almost never a consideration when analysing forest practices regulations. The more subtle difficulty questions the result of a regulation that merely *diminishes* the value of private property.

While much effort has been devoted to the effects of the federal constitutional protections against regulatory takings of private property, the emphasis is somewhat misplaced as the greater role that state constitutions may play is relatively ignored. State constitutions can present even greater obstacles to overreaching government forest regulations (Mortimer, 1999; Mortimer *et al.*, 2003). State constitutional assurances against diminutions or damage to the value of real property are far more rigorous than the federal standard.

While the question of regulatory taking by forest practices regulations has been largely settled since the late 1940s, changes in the type, form, and scope of state-level forestry regulations suggest that the legal situation is less than predictable. Recent case law indicates that courts may indeed treat as suspect laws that do not clearly remedy an identifiable public nuisance.[8] While forestry regulations may indeed be destined to increase in number (Siegel, 1997), as state laws and regulations evolve away from the abatement of nuisance (such as water quality and wildfire standards) towards more esoteric concerns

---

[8] *Boise Cascade* v. *State*, 164 Or. App 114, 991 P.2d. 563 (1999), *appeal denied* 331 Or. 244, 18 P.3d 1099 (2000), and *cert denied* 532 US 923 (2001). The court noted that knocking down a bird's nest, in this case an endangered spotted owl nest, had never been considered a public nuisance.

(such as wildlife habitat, aesthetics, and species distribution), the risk of judicial intervention increases.

For example, many of the southern states have recently proposed and/or enacted laws specifically designed to protect against unjustified government takings of forest or agricultural real property (USDA, 2002b). Regions of the US with economic and cultural connections to traditional resource industries, such as timber harvesting, reflect correspondingly strong private property rights traditions (Zheng, 1996).

19.7 Conclusions

There are, to this date, no programmatic national, regional or, in many cases, even state private forest policies, let alone explicit statutory consideration for forest biodiversity. None the less, concerns for forest biodiversity will continue to play an important role in scientific assessments of forest health. Popularization of concerns for preserving or enhancing forest biodiversity is also the mission of many US non-governmental organizations, such as the Nature Conservancy. However, the fact remains that heavy-handed or draconian steps to address biodiversity concerns by law or regulation will inevitably be met with vociferous resistance. Even if a particular regulation successfully withstands legal challenges, the tendency is for a particular success to polarize opposition elsewhere. While some suggest that property rights mechanisms might be modified to account for commodities such as biodiversity (Goldstein, 1998), the fact remains that restrictions on the use and management of private forestlands pose potentially serious constitutional and legal obstacles (Daughdrill and Zickert, 2001; Mortimer *et al.*, 2003). Alternatively, educating the various publics, citizens and policy-makers of the scientific, economic and social merits of forest biodiversity may set the stage for political acceptability. Alternative dispute mechanisms may be necessary (Thompson, 1993), and incentives for private landowners to manage their lands for biodiversity goals are also a more promising strategy, as erosion of constitutional guarantees (or of the US private property culture) is quite unlikely, particularly on the sole basis of forest biodiversity concerns.

References

Best, C. and Wayburn L.A. (2001) *America's Private Forest: Status and Stewardship.* Island Press, Washington, DC.

Birch, T.W. (1996) *Private Forest-land Owners of the United States, 1994.* USDA Forest Resources Bulletin NE-134, Northeastern Experimental Station.

Blackstone, W. (1979) *Commentaries on the Law of England. A Facsimile of the First Edition of 1765–1769.* University of Chicago Press, Chicago, Illinois.

California Division of Forestry and Fire Protection (2001) *Forest Practices Rules*. Sacramento, California.

Cubbage, F.W. (1997) The public interest in private forests: developing regulations and incentive. In: Kohm K.A. and Franklin, J.F. (eds) *Creating a Forestry for the 21st Century: the Science of Ecosystem Management*. Island Press, Washington, DC.

Dana, S.T. and Fairfax, S.K. (1980) *Forest and Range Policy*. McGraw-Hill, New York.

Daughdrill, B.E. and Zickert, K.M. (2001) Tree preservation ordinances: sacrificing private timber rights on the diminutive altar of public benefit. *Mercer Law Review* 52, 705–729.

Ellefson, P.V., Cheng, A.S., and Moulton, R.J. (1997) State forest practice regulatory programs: an approach to implementing ecosystem management on private forest lands in the United States. *Environmental Management* 21, 421–432.

Epstein, R.A. (1985) *Takings: Private Property and the Power of Eminent Domain*. Harvard University Press, Cambridge, Massachusetts.

Fitzsimmons, A.K. (1999) *Defending Illusions: Federal Protection of Ecosystems*. Rowman and Littlefield Publishers, Lanham, Maryland.

Goldstein, R.J. (1998) Green wood in the bundle of sticks: fitting environmental ethics and ecology into real property law. *Boston College Environmental Affairs Law Review* 25, 347–430.

Green, J.R. (1877) *History of the English People*. Macmillan, London.

Heinz Center (The H. John Heinz Center for Science, Economics and the Environment) (2002) *The State of the Nation's Ecosystems*. Cambridge University Press, Cambridge, UK.

Hickman, C.A. and Hickman, M.R. (1997) Legal limitations on governmental regulation of private forestry in the United States. In: Schmithüsen, F. and Siegel, W.C. (eds) *Developments in Forest and Environmental Law Influencing Natural Resource Management and Forestry Practices in the United States of America and Canada*. IUFRO World Series Vol. 7. Vienna.

Laitos, J.G. (2002) *Natural Resources Law*. West Group, St Paul, Minnesota.

Marcot, B.G. (1997) Biodiversity of old forests of the west: a lesson from our elders. In: Kohm, K.A. and Franklin, J.F. (eds) *Creating a Forestry for the 21st Century: the Science of Ecosystem Management*. Island Press, Washington, DC.

Meltz, R., Merriam, D.H., and Frank, R.M. (1999) *The Takings Issue: Constitutional Limits on Land Use Control and Environmental Regulation*. Island Press, Washington, DC.

Montana Department of Natural Resources and Conservation (2001) *Biodiversity and Old-Growth Management Standards*. Helena, Montana.

Mortimer, M.J. (1999) Condemnation without compensation: how environmental eminent domain diminishes the value of Montana's School Trust Lands. *Dickinson Journal of Environmental Law & Policy* 8, 243–271.

Mortimer, M.J., Haney, H.L. Jr and Spink, J.J. (2003) When worlds collide: science and policy at odds in the regulation of Virginia's private forests. *Journal of Natural Resources and Environmental Law* 17(1), 1–26.

National Research Council (1998) *Forested landscapes in perspective*. National Academy Press, Washington, DC.

Pregitzer, K.S., Goeble, P.C. and Wigley, T.B. (2001) Evaluating forestland classification schemes as tools for maintaining biodiversity. *Journal of Forestry* 99, 33–40.

Siegel, W.C. (1997) Legislative regulation of private forestry practices in the United States – recent trends in the United States. In: Schmithüsen, F. and Siegel, W.C. (eds) *Developments in Forest and Environmental Law Influencing Natural Resource Management and Forestry Practices in the United States of America and Canada*. IUFRO World Series Vol. 7, Vienna.

Society of American Foresters (2002) *Public Regulation of Private Forest Practices: A Position Statement of the Society of American Foresters*. Bethesda, Maryland.

Souder, J.A. and Fairfax, S.K. (1996) *State Trust Land: History, Management, and Sustainable Uses*. University of Kansas Press, Lawrence, Kansas.

Thomas, J.W. (2000) What now? From a former Chief of the Forest Service. In: Sedjo, R.A. (ed.) *A Vision for the U.S. Forest Service*. Resources for the Future, Washington, DC.

Thompson, R.P. (1993) Compensated takings and negotiated solutions: new hope for a balanced policy. *Journal of Forestry* 91, 14–18.

USDA (US Department of Agriculture, Forest Service) (1996) *Private Forest-land Owners of the United States, 1994*. Resource Bulletin NE-134. Northeastern Forest Experiment Station, USA.

USDA (US Department of Agriculture, Forest Service) (1997) *Forest Statistics of the United States*. Washington, DC.

USDA (US Department of Agriculture, Forest Service) (2002a) *2003 Report on Sustainable Forests*. Draft Report. Washington, DC.

USDA (US Department of Agriculture, Forest Service) (2002b) *Southern Forest Resource Assessment*. Southern Research Station, Asheville, North Carolina.

Utah Code (2001) Title 65A Chapter 8a Utah Forest Practice Act, USA.

Warren, H.D. (1950) Constitutionality of reforestation or forest conservation legislation. *American Law Reports 2d*. 13, 1095.

Washington Department of Natural Resources (2000) *Forest Practices Rule Book*. Olympia, Washington, DC.

Wiersum, K.F. (1995) 200 years of sustainability in forestry: lessons from history. *Environmental Management* 19, 321–329.

Zeide, B. (2001) Resolving contradictions in forestry: back to science. *The Forestry Chronicle* 77, 973–981.

Zheng, D. (1996) State property rights: what, where, and how? *Journal of Forestry* 94, 10–15.

Index

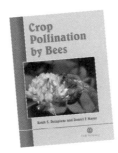
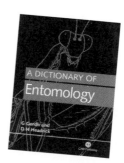
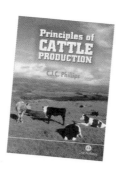